T0213455

What Every Engineer Should Know About the Internet of Things

What Every Engineer Should Know

Series Editor:
Phillip A. Laplante
Pennsylvania State University

What Every Engineer Should Know About Developing Real-Time
Embedded Products
Kim R. Fowler

What Every Engineer Should Know About Business Communication
John X. Wang

What Every Engineer Should Know About Career Management
Mike Ficco

What Every Engineer Should Know About Starting a High-Tech
Business Venture
Eric Koester

What Every Engineer Should Know About MATLAB® and Simulink®
Adrian B. Biran

Green Entrepreneur Handbook: The Guide to Building and Growing a
Green and Clean Business
Eric Koester

Technical Writing: A Practical Guide for Engineers and Scientists
Phillip A. Laplante

What Every Engineer Should Know About Cyber Security and Digital Forensics
Joanna F. DeFranco

What Every Engineer Should Know About Modeling and Simulation
Raymond J. Madachy and Daniel Houston

What Every Engineer Should Know About Excel, Second Edition
J.P. Holman and Blake K. Holman

Technical Writing: A Practical Guide for Engineers, Scientists, and
Nontechnical Professionals, Second Edition
Phillip A. Laplante

What Every Engineer Should Know About the Internet of Things
Joanna F. DeFranco and Mohamad Kassab

For more information about this series, please visit: www.routledge.com/What-Every-Engineer-Should-Know/book-series/CRCWEESK

What Every Engineer Should Know About the Internet of Things

Joanna F. DeFranco and Mohamad Kassab

CRC Press
Taylor & Francis Group
Boca Raton London New York

CRC Press is an imprint of the
Taylor & Francis Group, an **informa** business

First edition published 2022
by CRC Press
6000 Broken Sound Parkway NW, Suite 300, Boca Raton, FL 33487-2742

and by CRC Press
2 Park Square, Milton Park, Abingdon, Oxon, OX14 4RN

© 2022 Joanna F. DeFranco and Mohamad Kassab

CRC Press is an imprint of Taylor & Francis Group, LLC

Library of Congress Cataloging-in-Publication Data
Names: DeFranco, Joanna F., author. | Kassab, Mohamad, author.
Title: What every engineer should know about the internet of things /
Joanna F. DeFranco and Mohamad Kassab.
Description: Boca Raton, FL : CRC Press, [2022] | Series: What every
engineer should know | Includes bibliographical references and index.
Identifiers: LCCN 2021022598 (print) | LCCN 2021022599 (ebook) |
ISBN 9780367462727 (hbk) | ISBN 9780367858780 (pbk) |
ISBN 9781003027799 (ebk)
Subjects: LCSH: Internet of things.
Classification: LCC TK5105.8857 .D475 2022 (print) |
LCC TK5105.8857 (ebook) | DDC 004.67/8—dc23
LC record available at https://lccn.loc.gov/2021022598
LC ebook record available at https://lccn.loc.gov/2021022599

ISBN: 978-0-367-46272-7 (hbk)
ISBN: 978-0-367-85878-0 (pbk)
ISBN: 978-1-003-02779-9 (ebk)

DOI: 10.1201/9781003027799

Typeset in Times
by codeMantra

Contents

Preface

In 2022, worldwide Internet of Things (IoT) spending will exceed $1 trillion. By 2025, there will be 75 billion connected smart devices. IoT products and cyber-physical systems (CPS) are being utilized in almost every discipline. In addition, smart products are becoming more advanced to improve our health and comfort. There are IoT applications for healthcare that not only provide information to providers to assist decision-making that is based on real-time data, but also can improve a patient's treatment regime for diseases that require around the clock care. There are products that sense, learn, and react to user preferences that are gaining popularity. IoT is also making our cities, government, and homes run more efficiently.

IoT devices have a ubiquitous nature; they are being integrated into every domain. With this increase, it is of utmost importance to understand and communicate how this technology, with its distinguished system functions such as sensing and decision-making, can support and challenge all interrelated actors as well as all involved assets of many domains.

There are three main goals of this book. The first goal is to provide an introduction to the IoT that can be understood and appreciated by every engineering discipline. We include chapters that highlight the top applications of IoT (Smart Homes, Smart Healthcare, Smart Cities, and Smart Education). The second goal is to provide information to engineers who have an interest in design, with chapters on IoT requirements and architecture, IoT and blockchain, and the National Institute of Standards and Technology (NIST) definitions of Networks of 'Things' and their white paper on IoT Trust Concerns. The third goal is to provide insight to engineering educators on educating the next generation of engineers who are needed to build IoT systems. We hope the book will give you a new appreciation for IoT and its many applications, as it will probably be integrated into almost every part of our lives sometime soon.

Authors

Dr. Joanna F. DeFranco is an associate professor of software engineering and a member of the graduate faculty at The Pennsylvania State University. Before her academic career, she spent many years as a software engineer for government and industry. Notable engineering experiences during this period include traveling the world on Naval scientific ships while developing and testing ocean mapping software for the Naval Air Development Center (NADC) and developing software for cable headend devices at Motorola. She has published many journal articles and conference proceedings on software security, IoT, blockchain, and software engineering. She authored another book in this series on Cyber Security and Digital Forensics and has coauthored a book on project management.

In addition to academia, Dr. DeFranco is currently a researcher for the National Institute of Standards and Technology (NIST), the Computing Fundamentals area editor for IEEE Computer, and the IoT column editor of *IEEE Computer* magazine. As for her scholarly credentials, she earned a BS in electrical engineering and math from Penn State University, an MS in computer engineering from Villanova University, and a PhD in computer and information science from New Jersey Institute of Technology.

Dr. Mohamad Kassab is an associate research professor in software engineering at Pennsylvania State University, and he earned his PhD and MS degrees in software engineering from Concordia University in Montreal, Canada. Previously, Dr. Kassab has been a postdoctoral researcher at ETS School of Advanced Technology in Montreal and visiting scholar at Carnegie Mellon University. Dr. Kassab's research interests include requirements engineering, system architecture, software quality and measurements, blockchain, and the Internet of Things. He has published extensively in software engineering books and journals. With more than 18 years of industrial experience, he has worked in various industrial roles which include business unit manager at Soramitsu, senior quality engineer at SAP, senior associate at Morgan Stanley, senior quality assurance specialist at NOKIA, and senior software developer at Positron Safety Systems.

Acknowledgments

Many people provided invaluable support and assistance in various ways during the writing of this manuscript. We want to take this opportunity to thank the following people:

- Dr. Phillip Laplante, for his mentoring as well as his collaborations in this book.
- Dr. Jeff Voas, for his contributions of the NIST work included in this book as well as his contributions to the chapter on IoT Education.
- Dr. Nancy Laplante, for her contribution on IoT healthcare.
- Mr. Michael Hutchinson, for his contribution on IoT healthcare.
- Drs. Hyunji Chung and Jungheum Park, for their contributions on smart city security.
- Drs. Valdemar Vicente Graciano Neto and Renato F. Bulcão-Neto, for his contribution on smart cities.
- Drs. Giuseppe Destefanis, Lodovica Marchesi, Michele Marchesi, Marco Ortu, and Roberto Tonelli, for their contributions on IoT and blockchain.
- Mr. Mohamad Murywed, for his contribution on Smart Cities - Energy.
- Ms. Nina DeFranco Tommarello, for her contribution on the smart garden project.

ERRORS

Despite our best effort as well as the effort of the reviewers and the publisher, there may be errors in this book. If errors are found, please report them to jfd104@psu.edu or muk36@psu.edu.

1 Internet of Things Defined[1]

1.1 INTRODUCTION

If you ask the average person to explain "Internet of Things" (IoT[2]), they probably will respond that they never heard of IoT or have no idea how to explain it. If they make a guess, they may tell you that IoT are 'things' that are 'smart'. Most likely they do not understand what makes the device smart. In fact, the term 'smart' seems to also have become a marketing term on products that don't warrant that description. Consider outdoor lights that can be operated from your phone. Those lights are not smart unless they also sense something such as motion and then send a notification to your phone. In other words, 'smart' should refer to a device that meets certain criteria such as embedded sensors and analytical software. Specifically, if along with the sensing, the device can also collect and analyze data, is connected to a wired or wireless network, and then performs some type of action based on data analysis, it can be called an IoT device or a 'smart' device. There are many inconsistent and inaccurate definitions of IoT; thus, one goal of this book is to provide clarity and accuracy in understanding IoT and its applications. In addition, as engineers, we need to understand everything about IoT – especially the benefits, risks, and how to continually improve the security, safety, privacy, scalability, performance, interoperability, and usability of our IoT systems.

1.2 IoT DEFINED

Wikipedia defines IoT as "the network of physical objects – a.k.a. 'things' that are embedded with sensors, software, and other technologies for the purpose of connecting and exchanging data with other devices and systems over the Internet". The IoT concept emerged from the RFID (radio-frequency identification) community and initially focused on the ability to track location and provide a status for a physical object (Greer et al., 2019). This greatly improved the visibility in supply chains.

The term "Internet of Things" has been around since 1999. Kevin Ashton, co-founder of MIT's Auto-ID Labs, coined the term IoT to refer to a combination of real and virtual information technology that are connected through automatic identification technologies (e.g., RFID), location systems (global positioning system (GPS)), sensors (e.g., temperature sensor), and actuators (e.g., electric motor). There are many types of sensors, such as RFID tags, GPS, and accelerometers. 'Things' could be

---

[1] Portions of this chapter were contributed by Dr. Phillip Laplante.
[2] We refer to "Internet of Things", "IoT", or "the Internet of Things" throughout this book. These forms are all equivalent in meaning.

DOI: 10.1201/9781003027799-1

cameras, social media feeds, and cyber and physical devices – anything that can produce data. The things can be wired or wireless – although most are wireless due to the scale of IoT and the limitations of wired infrastructures.

The National Institute of Standards and Technology (NIST), Special Publication (800-183), describes the set of building blocks that "govern the operation, trustworthiness, and lifecycle of IoT" (Voas, 2016). NIST is a federal agency whose central mission is to support innovation. In order to design and implement IoT innovations, it is extremely important to understand the underlying and foundational science behind IoT; thus, NIST described five primitives (sensor, aggregator, communication channel, external utility (eUtility), and decision trigger) that may be used to build an IoT system. This is included in Chapter 2 of this book.

What do the "things" do? They capture data and communicate/transmit the data to algorithms (software) that may trigger a decision or perform some action (e.g., an actuator). IoT has enabled many new applications, helped to expand domains such as big data, business and data analytics, and artificial intelligence. Most importantly, the IoT has an impact on quality of life for many.

1.3   IoT ECOSYSTEMS

The IoT permeates many aspects of our lives. Part of this trend is due to the falling prices of RFID sensors, microelectromechanical devices (e.g., accelerometers, gyroscopes, pressure sensors), Wi-Fi routers, as well as the ubiquity of mobile devices and the decreasing size and increasing capabilities of microcontrollers. IoT has improved or will improve the way our homes, classrooms, shopping, manufacturing, and health service systems function. These advances have created many IoT ecosystems (a connection of various kinds of devices that sense and analyze data and then communicate with each other over a network) such as industrial internets, cyber-physical systems (CPS), sensor networks, clouds, and the Internet. IoT ecosystems can be found (and will appear) in virtually every important application domain including critical infrastructure, entertainment, and every form of pedestrian convenience in the public space and in the home. IoT ecosystems will be used to improve sustainability and accessibility in these spaces as well.

IoT has also made an enormous impact on our critical infrastructure systems. When the "things" are networked physical systems, it is called a cyber-physical system (CPS). Critical infrastructure systems fall in the CPS category. A system is deemed critical when the assets, systems, and networks are considered so vital that if compromised, would have a debilitating effect on US security, economic security, and national public health or safety. IoT allows these physical devices to connect, network, exchange data, and provide real-time decision support. The Cybersecurity Infrastructure Security Agency (CISA – https://www.cisa.gov/) names and defines the 16 critical infrastructure sectors below:

- *Chemical Sector*: The part of the economy that manufactures, stores, uses, and transports potentially dangerous chemicals.
- *Commercial Facilities Sector*: A diverse range of sites that draw large crowds of people for shopping, business, entertainment, or lodging.

- *Communications Sector*: The underlying operations for businesses, public safety, and government.
- *Critical Manufacturing Sector*: Primary metals, machinery, electrical equipment, appliance, transportation, and component manufacturing.
- *Dams Sector*: Navigation locks, levees, hurricane barriers, water retention facilities, etc.
- *Defense Industrial Base Sector*: The maintenance of military weapons systems, subsystems, and components or parts to meet US military requirements.
- *Emergency Services Sector*: Agencies that prevent/prepare/respond/recover a range of emergency services.
- *Energy Sector*: The energy supply.
- *Financial Services Sector*: Depository institutions, providers of investment products, insurance companies, other credit and financing organizations, and the providers of the critical financial utilities and services.
- *Food and Agriculture Sector*: Food manufacturing/processing/storage facilities – also has critical dependency on other critical infrastructure (water, transportation, energy, chemical).
- *Government Facilities Sector*: A wide variety of US-owned facilities.
- *Healthcare and Public Health Sector*: Public and private healthcare organizations that respond and provide recovery across all other sectors in the event of a natural or man-made disaster.
- *Information Technology Sector*: The virtual and distributed functions that produce and provide hardware, software, and information technology systems and services, and the Internet.
- *Nuclear Reactors, Materials, and Waste Sector*: The power reactors that provide electricity to millions of Americans, to the medical isotopes used to treat cancer patients, the nuclear reactors, materials, and waste sector.
- *Transportation Systems Sector*: This sector includes aviation, highway, maritime, mass transit, passenger rail, pipeline, freight, and postal.
- *Water and Wastewater Systems Sector*: Public drinking water and wastewater treatment systems.

1.4 IoT APPLICATIONS

IoT is slowly pervading every part of our lives.[3] Consider the *Amazon Go* app used at an Amazon physical store. Using a mobile app and RFID tags, the customer takes an item off a shelf and an instantaneous charge is posted to their account (and the store inventory data is decreased!). If the customer changes their mind, they can put the item back on the shelf to credit their account and the store inventory. Using computer vision, deep learning, and RFID, a customer can walk in the store, take an item off the shelf, put the item in their bag, and walk out of the store. The time saved is amazing. The remainder of this section will cover IoT in the most prevalent domains.

---

[3] Some of this section has been excerpted from Joanna F. DeFranco and Michael Hutchinson, "Understanding Smart Medical Devices," *Computer*, May 2021.

Healthcare: IoT is transforming the healthcare domain in very beneficial ways. IoT tools such as those needed for healthcare monitoring and compliance have been and are being developed. Smart healthcare is any health service using technology to collect and store information that can be dynamically accessed by the patient, provider, or another device. A great example is a device that helps prevent a healthcare-associated infection (HAI). On any given day, 1 out of 31 hospital patients acquires an HAI (Centers for Disease Control). Several studies reported that a simple and straightforward process where a provider takes only a few seconds to clean their hands with an alcohol-based hand rub helps prevent HAIs and save lives, reduces morbidity, and minimizes healthcare costs (Haque et al., 2018). However, the hand cleaning was not consistent. A tool that automatically detects whether staff members wash their hands as they enter and exit a patient's room was developed. In addition, this tool collects and provides the compliance data needed for infection preventionists. Here is how it works: The soap/gel dispensers have embedded software that detects the Wi-Fi badge worn by the healthcare professional – then the hand hygiene data is reported. An individual report will include the hand hygiene transactions (when, where, who) of the healthcare professional. The managers can then determine hygiene gaps that need to be addressed to improve the hygiene process. Other IoT healthcare domain products track the location of mobile testing equipment and systems for the pharmaceutical and biotech industry to remotely monitor and provide notification for network outages and when the temperature, humidity, and CO_2 levels exceed desired parameters.

Another significant smart healthcare category are wearable devices. Some are used for location monitoring to ensure the safety of seniors and babies during hospital stays. The best example of a patient-wearable medical device is a closed-loop insulin delivery system (a.k.a. artificial pancreas) for a person with type 1 diabetes (T1D). T1D is a chronic incurable autoimmune disease where the person is insulin *dependent*. The challenge is that the amount of insulin a patient needs varies depending on many factors.

Without devices, type 1 patients use needle therapy, and they require two types of insulin: long-acting (also called "basal") and fast-acting (also called "bolus"). A closed-loop insulin delivery system uses an IoT architecture to control insulin delivery. To appreciate this type of IoT device, and the risk of automating care, some knowledge of T1D care is necessary. The greatest risk is that if too much insulin is dosed, the patient may experience low blood sugar – which can be life-threatening. If the patient does not receive enough insulin, the risk is vital organ damage as well as an increased risk of other autoimmune diseases. The challenge is that the amount of insulin each patient needs varies and depends on many factors. To keep blood sugar level in a normal range, the dosed amount is dependent on the food (amount and type), body size, hormone levels, activity, current health status, time of day, current amount of insulin in the body, and even the weather at times (https://www.diabetes.co.uk/).

An insulin pump still requires manual input to dose when food is consumed or to correct high blood sugar. In other words, the user needs to manually enter the blood sugar reading from a continuous glucoses monitor (CGM) (or a finger prick) into the pump.

Where does the IoT come into play? Both devices, the CGM and insulin pump, are worn by the patient, and the CGM wirelessly sends the glucose level directly to the pump. That connection closes the loop to automatically deliver *some* of the insulin needed. Specifically, the CGM reading is received by the pump, and the pump uses an algorithm to detect when the glucose level is rising or falling. It is important to understand that the interstitial CGM reading lags behind the actual blood glucose reading because it takes time for the glucose level to reach the interstitial fluid. By interpreting the steepness of the slope, the pump algorithms account for the lag. Depending on the glucose trend, the pump will automatically dispense either bolus insulin to address a spike in blood sugar or scale back on the basal insulin when blood sugar level is dropping. The patient still needs to dose bolus insulin for the food consumed (insulin amount varies based on food carbohydrates). Thus, the closed-loop system still requires user manual entry when food is consumed. The exciting part is that even more advanced IoT devices to care for T1D patients are being developed and tested to address not only meal reporting but also maintaining a normal blood sugar range after meals and/or activity. This is peace of mind for parents as some adolescent T1D patients frequently forget to dose insulin for food consumed (Palisaitis et al., 2021).

Education: The education industry is also integrating IoT into everything from inventory to student engagement to remote laboratories. Similar to the healthcare domain, there are IoT systems to manage and track academic resources such as portable projectors, laboratory equipment, and sports equipment. The collected data from tracking portable equipment can be used to determine equipment use patterns and trends. Other IoT educational devices use sensors to detect noise levels and manage environmental conditions (e.g., lighting/heat). The most innovative IoT systems in education are smart environments to support learning and improve the quality of education, for example, providing support to students with adapted learning resources by integrating/pushing content to a student that is based on the student's location and knowledge level. Another example is virtual laboratories to lower costs and engage students. Virtual laboratories not only provide accessibility for school districts that may have a resource problem but can also provide personalized experience for the student by using data analytics to monitor a student's progress and adjust the curriculum appropriately. This is covered in detail in Chapters 7 and 8.

Smart Homes: Who would not want a refrigerator that restocks itself? aside from one of the authors' mother, who LOVES the grocery store. This refrigerator does not exist *yet*; however, there is a beer refrigerator that restocks itself! The smart beer refrigerator senses low inventory as you remove beers from the device and automatically reorders beer from the vendor of your choice. Current smart home development falls into five categories: *security design and management* (e.g., security systems and management, device security, risk management, security architecture, application security, intrusion detection, encryption, authentication, and privacy); *product* (e.g., home comfort, gardening assistance, entertainment, applications to control household devices, home automation systems, smart house assistants, health notification/monitoring); *activity and behavior patterns* (e.g., identifying human activity and behavior patterns, models for planned behavior, and user location and discovery (ULD)); *power efficiency* (e.g., device/system power consumption/efficiency and energy optimization); and *systems design, simulation* (systems design, models, simulations, requirements, architectures, frameworks, cost models, performance-improving algorithms, and interoperability solutions). More on this topic can be found in Chapter 6.

Smart Cities: Smart cities is a vast topic that is discussed in a few chapters in this book. Smart cities use technology to improve traffic patterns, energy distribution, air quality, and more. The elements of a smart city can also increase the convenience and quality of life for its citizens. Information technology is integrated into many aspects of citizen interaction such as simplifying access to many of the city's services. The fields of healthcare, education, culture, and shopping can all be integrated into the core of a smart city to create an infrastructure that allows citizens to live more conveniently.

1.5   IoT CHALLENGES

Clearly IoT-enhanced systems have many benefits such as ubiquitous network connectivity, real-time response, enhanced situational awareness, and process optimization (CISA, 2019). The IoT system challenges are the same quality requirements that should be designed into any system. Here are a few examples that will be expanded upon in Chapter 12, the requirements chapter of the book:

- *Security*: Systems should be resilient to attacks (e.g., avoiding single points of failure); authenticate retrieved data; implement access control; maintain privacy.
- *Trust*: This is an involved topic as the IoT connections are complex and systems inherit a core set of trust concerns. Trust is very difficult to test and prove. This will be covered in Chapter 10, where the NIST Internal Report 8222 (NISTIR 8222 (DRAFT)) is included. This report details 17 technical trust-related concerns for individuals and organizations before and after IoT adoption (Voas et al., 2018).
- *Reliability*: IoT systems should perform as expected and should not terminate abnormally. Maintenance and repairs due to reliability issues need to be avoided.
- *Safety*: IoT applications should have parameters that stay in a specific range. A safety (and security) violation can cause severe harm to its operational environment (Zalewski, 2019). Consider the levels of safety concerns shown in Figure 1.1. Individuals are wearing medication dosing IoT devices, households have IoT security devices protecting the homeowners, smart cities have automated systems (e.g., traffic, power/energy distribution), and industry has critical infrastructure.
- *Scalability*: The system should be able to grow with minimal interruption. Address traffic and loading issues – especially in private clouds.
- *Vulnerability*: The system should be tested for possible attacks. With the increased connectivity in IoT systems, there is an increased attack surface. Vulnerabilities can be inadvertently created or deliberately planted.

One of the most important challenges for IoT systems is the inadvertent creation of ad hoc networks. That is, due to the high level of connectivity using wireless devices, an IoT can accidentally, or through malicious intent, make a connection to another IoT network in range, causing undesirable consequences. For example, imagine if a

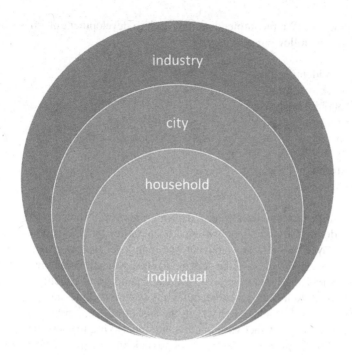

FIGURE 1.1 Layered prospective of IoT safety concerns. (Adapted from Zalewski, 2019.)

nearby IoT-enabled vehicle connects to a nearby network and causes interference in an IoT-connected critical infrastructure system, such as a power plant; therefore, a well-conceived and structured development approach is needed to avoid creation of unwanted IoT system interactions.

1.6 IoT DEVELOPMENT APPROACH

IoT system development requires a comprehensive and reliable approach that encompasses the development processes (e.g., requirements engineering, design, construction, and testing), the people involved in the development (i.e., qualification through certification and licensure), and the use of appropriately certified tools. Application domain governance entities set the standards for the appropriate qualifications, certifications, and standards involved in the development of these systems. For example, in the United States, there are governing agencies for avionics (the Federal Aviation Administration), medical devices (Federal Drug Administration), and for transportation of vehicles and infrastructure (Department of Transportation), to name just a few. For other types of systems, other governing bodies, professional organizations, and standards boards help to ensure process discipline.

Whether an IoT system is a critical one being developed under the discipline of an appropriate regulating or guiding agency or an informal one with no particular rules, it is imperative that the system be built with a structured, disciplined approach. Using such process discipline can also promote standardization, reuse, interoperability, and best practices. We recommend an approach to development of IoT systems that encompasses the following activities:

1. Mission identification
2. Goal identification
3. Stakeholder analysis
4. Standards identification
5. Requirements writing/agreement/verification/validation
6. Design
7. Construction
8. Testing

Mission identification involves clearly defining the overall intent of the system, the application domain or domains in which it is operating, and the appropriate governing entities. From the mission statement, the high-level system goals can be identified. From these goals, the mission statement, and application domain identification, a set of stakeholders (direct users, indirect users, people and systems impacted, etc.) can be identified. Relevant stakeholders for any system include manufacturers, owners, and users of IoT-enabled artifacts in the system under consideration and those which could possibly interact with that system. Identifying the stakeholders permits an analysis of their concerns and sources of possible interactions. Then approaches to eliminate or mitigate these unwanted interactions can be undertaken.

Next, all relevant standards governing the application domain and the devices and systems used need to be identified. Any applicable laws, regulations, standards, etc., e.g., those concerning wireless communications, need to be identified. All of these steps are preparation for the development of a comprehensive System Requirements Specification (SRS), which is discussed in detail in Chapter 12.

Design of the system follows, ideally using structured approach that is somewhat unique for IoT systems (e.g., identifying the IoT elements and primitives). Chapter 10 covers many of these ideas.

Finally, systems testing (verification and validation levels) must be conducted – although relevant testing and validation activities must be conducted throughout the systems' life cycle.

It would take many texts to completely cover the various life cycle activities for the development and deployment of an IoT system. The foregoing discussion is simply an overview of the main activities involved.

FURTHER READING

Centers for Disease Control, "HAI Data." https://www.cdc.gov/hai/data/index.html, October 5, 2018 retrieved 10/13/2019.

CISA, "The Internet of Things: impact on public safety communications," March 2019. https://www.cisa.gov/sites/default/files/publications/CISA%20IoT%20White%20 Paper_3.6.19%20-%20FINAL.pdf.

Greer, C., Burns, M., Wollman, D., Griffor, E., "Cyber-physical systems and Internet of Things," NIST Special Publication 1900-202, March 2019. https://nvlpubs.nist.gov/nistpubs/SpecialPublications/NIST.SP.1900-202.pdf.

Haque, M., Sartelli, M., McKimm, J., Bakar, M., "Health care-associated infections – an overview," *Infection and Drug Resistance*, vol. 11, 2018, pp. 2321–2333.

Palisaitis, E., Fathi, A., Oettingen, J., Haidar, A., Legault, L., "A meal detection algorithm for the artificial pancreas: a randomized controlled clinical trial in adolescents with type 1 diabetes," *Canadian Journal of Diabetes*, vol. 44, 2021, pp. 604–606.

Voas, J., "Networks of 'Things'," Draft NIST Special Publication 800-183, July 2016. doi:10.6028/NIST.SP.800-183.

Voas, J., Kuhn, R, Laplante, P., Applebaum, S., "Internet of Things (IoT) trust concerns," NIST Cybersecurity White Paper, 2018. https://csrc.nist.gov/publications/detail/white-paper/2018/10/17/iot-trust-concerns/draft.

Zalewski, J., "IoT safety: state of the art," *IT Professional*, vol. 21, no. 1, January/February 2019, pp. 16–20.

2 Networks of 'Things'[1]

2.1 INTRODUCTION

This chapter contains an essential National Institute of Standards and Technology (NIST) standard (SP 800-183) that describes the building blocks of IoT (Voas, 2016). A standard is a document that provides the definitions needed for technologies to be built and work together seamlessly. NIST is an organization with a central mission of promoting innovation and industrial competitiveness. Founded in 1901, NIST is a nonregulatory federal agency within the U.S. Department of Commerce. NIST innovations are based on measurement science, standards, and technology to enhance economic security and improve our quality of life.

The five core primitives belonging to the Internet of Things (IoT) systems, and to most distributed systems, are presented in this chapter. System primitives allow formalisms, reasoning, simulations, and reliability and security risk trade-offs to be formulated and argued. These primitives apply well to systems with large amounts of data, scalability concerns, heterogeneity concerns, temporal concerns, and elements of unknown pedigree with possible nefarious intent. These primitives are the basic building blocks for networks of 'things' (NoT), including the IoT. This offers an underlying and foundational understanding of IoT based on the realization that IoT involves sensing, computing, communication, and actuation. The material presented here is generic to all distributed systems that employ IoT technologies (i.e., 'things' and networks).

2.2 NETWORKS OF 'THINGS'

From agriculture, to manufacturing, to smart homes, to healthcare, and beyond, there is value in having numerous sensory devices connected to larger infrastructures. This technology advance acknowledges the reality that human society is moving towards 'smart' and 'smarter' systems. The rapid advances in computer science, software engineering, systems engineering, networking, sensing, communication, and artificial intelligence (AI) are converging. The tethering factor is data.

There is no formal, analytic, or even descriptive set of the building blocks that govern the operation, trustworthiness, and life cycle of IoT. A composability model and vocabulary that defines principles common to most, if not all networks of things, is needed to address the question: "what is the science, if any, underlying IoT?" NIST SP 800-183 offers an underlying and foundational science to IoT based on a belief that IoT involves *sensing, computing, communication,* and *actuation.*

Two acronyms will be used in this chapter, IoT and NoT, extensively and interchangeably – the relationship between NoT and IoT is subtle. IoT is an instantiation of a NoT, more specifically, IoT has its 'things' tethered to the Internet. A different

---

[1] This chapter was contributed by Dr. Jeffrey Voas.

DOI: 10.1201/9781003027799-2

type of NoT could be a local area network (LAN), with none of its 'things' connected to the Internet. Social media networks, sensor networks, and the Industrial Internet are all variants of NoTs. This differentiation in terminology provides ease in separating out use cases from varying vertical and quality domains (e.g., transportation, medical, financial, agricultural, safety-critical, security-critical, performance-critical, high assurance, to name a few). That is useful since there is no singular IoT, and it is meaningless to speak of comparing one IoT to another.

Primitives are building blocks that offer the possibility of an answer to the aforementioned questions and statements by allowing comparisons between NoTs. We use the term "primitive" to represent smaller pieces from which larger blocks or systems can be built. For example, in software coding, primitives typically include the arithmetic and logical operations (plus, minus, and, or, etc.). In this document, we do not employ the restriction that primitives cannot be developed or derived from something else; this is often a common, paraphrased definition for "primitive", but it is not employed here.

This model does not specify a definition for what is or is not a 'thing'. Instead, we consider that each primitive injects a behavior representing that 'thing' into a NoT's workflow and dataflow. 'Things' can occur in physical space or virtual space. In physical space, consider humans, vehicles, residences, computers, switches, routers, smart devices, road networks, office buildings, etc. In virtual space, consider software, social media threads, files, data streams, virtual machines, virtual networks, etc. More ideas concerning 'things' are presented later in this chapter.

Primitives offer a unifying vocabulary that allows for composition and information exchange among differently purposed networks. They offer clarity regarding concerns that are subtle, including interoperability, composability, and continuously binding assets that come and go on the fly. Because no simple, actionable, and universally accepted definition for IoT exists, the model and vocabulary proposed here reveals underlying foundations of the IoT, i.e., they expose the ingredients that can express how the IoT *behaves*, without defining IoT. This offers insights into issues specific to trust.

Further, we employ a paraphrased, general definition for a *distributed system*: a software system in which components located on networked computers communicate and coordinate their actions by passing messages. The components interact with each other in order to achieve a common goal.[2] NoTs satisfy this definition. Thus, we consider IoT to be one type of a NoT and a NoT to be one type of a distributed system.

2.3   THE PRIMITIVES

The *primitives* of a NoT are (1) sensor, (2) aggregator, (3) communication channel, (4) external utility (*e*Utility), and (5) decision trigger. There may be some NoTs that do not contain all of these, but that will be rare.

Each primitive, along with its definition, assumptions, properties, and role, is presented. We employ a dataflow model, captured as a sequence of four figures in this

---

[2] George Coulouris et al., *Distributed Systems: Concepts and Design*, 5th ed. (Boston: Addison-Wesley, 2011).

document, to illustrate how primitives, when composed in a certain manner, could impact a confidence in trustworthiness of NoTs. Although this model may seem overly abstract at first glance, its simplicity offers a certain elegance by not overcomplicating IoT's small handful of building blocks.

2.3.1 Primitive #1: Sensor

A *sensor* is an electronic utility that measures physical properties such as temperature, acceleration, weight, sound, location, presence, and identity. All sensors employ mechanical, electrical, chemical, optical, or other effects at an interface to a controlled process or open environment. Basic properties, assumptions, recommendations, and general statements about sensor include the following:

1. Sensors are physical; some may have an Internet access capability.
2. Sensor output is data; in our writings, $s_1 \rightarrow d_1$ means that sensor 1 has produced a piece of data that is numbered 1. Likewise, $s_2 \rightarrow d_2$ means that sensor 2 has produced a piece of data that is numbered 2. ($d_1, d_2, d_3, \ldots d_n$ will likely be digital data.) Analog sensors such as microphones and voltmeters are counterexamples.
3. A sensor may also transmit device identification information, such as via Radio Frequency Identification (RFID).[3]
4. Sensors may have an identity or have the identity of the 'thing' to which they are attached.
5. Sensors may have little or no software functionality and computing power; more advanced sensors may have software functionality and computing power.
6. Sensors may be heterogeneous, from different manufacturers, and collect data, with varying levels of data integrity.
7. Sensors may be associated with fixed geographic locations or may be mobile.
8. Sensors may provide surveillance. Cameras and microphones are sensors.
9. Sensors may have an owner(s) who will have control of the data their sensors collect, who is allowed to access it, and when.
10. Sensors will have pedigree – geographic locations of origin and manufacturers. Pedigree may be unknown and suspect.
11. Sensors may be cheap, disposable, and susceptible to wear out over time.
12. There may be differentials in sensor security, safety, and reliability, e.g., between consumer grade, military grade, and industrial grade.
13. Sensors may return no data, totally flawed data, partially flawed data, or correct and acceptable data. Sensors may fail completely or intermittently. They may lose sensitivity or calibration.

---

[3] RFID is an automatic identification method that stores and remotely retrieves data via a RFID tag or transponder. A RFID programmer encodes information onto a tiny microchip within a thin RFID inlay. In supply chain applications, these inlays typically are embedded in a tag that looks similar to pressure sensitive labels. RFID inlays can be applied to a wide variety of NoTs. RFID technology offers security and reliability features that enhance trustworthiness in IoT ecosystems.

14. Sensors are expected to return data in certain ranges, e.g., [1 … 100]. When ranges are violated, rules may be needed on whether to turn control over to a human or machine when ignoring out-of-bounds data is inappropriate.
15. Sensors may be disposable or serviceable in terms of calibration, sensitivity, or other forms of refresh. Complex and expensive sensors may be repaired instead of replaced.
16. Sensors may be powered in a variety of ways including alternating current (AC), solar, wind, battery, or passively via radio waves.
17. Sensors may be acquired off the shelf or built to specification.
18. Sensors acquire data that can be event-driven, driven by manual input, command-driven, or released at predefined times.
19. Sensors may have a level of data integrity ascribed (see Section 2.2).
20. Sensors may have their data encrypted to void some security concerns.
21. Sensors should have the capability to be authenticated as genuine.
22. Sensor data may be sent and communicated to multiple NoTs. A sensor may have multiple recipients of its data. Sensor data may be leased to one or more NoTs.
23. The frequency with which sensors release data impacts the data's currency and relevance. Sensors may return valid but stale data. Sensor data may be 'at rest' for long periods of time.
24. A sensor's precision may determine how much information is provided. *Uncertainty* of sensor data should be considered.
25. Sensors may transmit data about the "health" of a system, such as is done in prognostics and health management (PHM).
26. In this document and model, we do not classify humans as sensors; humans are classified as an *e*Utility, our fourth primitive (see Section 2.4), and/or classified as our fourth element, owner (see Section 2.3). When classified as an *e*Utility, humans can still act in a sensor-like role by manually feeding data into a NoT's workflow and dataflow.
27. Humans can influence sensor performance through failure to follow policy, sensor misplacement, etc. (or their positive analogs). Humans are potential contributors to sensor failures.
28. Security is a concern for sensors if they or their data is tampered with, stolen, deleted, dropped, or transmitted insecurely so it can be accessed by unauthorized parties. Building security into specific sensors may or may not be necessary based on the overall system design.
29. Reliability is a concern for sensors.

2.3.2 PRIMITIVE #2: AGGREGATOR

An *aggregator* is a software implementation based on mathematical function(s) that transforms groups of *raw* data into *intermediate, aggregated* data. Raw data can come from any source. Aggregators help in managing 'big' data. Basic properties, assumptions, recommendations, and general statements about aggregator include the following:

1. Aggregators may be virtual due the benefit of changing implementations quickly and increased malleability. A situation may exist where aggregators are physically manufactured, e.g., a field-programmable gate array (FPGA) or hard-coded aggregator that is not programmable. Aggregators may also act in a similar way as *n*-version voters.

2. Aggregators require computing horsepower; however, this assumption can be relaxed by changing the definition and assumption of virtual to physical, e.g. firmware, microcontroller, or microprocessor. For example, aggregators could execute on faster hardware such as a smartphone. Aggregators will likely use weights (see Section 2.2) to compute intermediate, aggregated data.

3. Aggregators have two actors for consolidating large volumes of data into lesser amounts: clusters (see Section 2.1) and weights (see Section 2.2). Aggregators process *big data* concerns within NoTs, and to satisfy this role, computational "performance enhancing" technologies will be needed. This is the only primitive with actors.

4. Sensors may communicate directly with other sensors, and thus act in some situations quite similar to aggregators.

5. Intermediate, aggregated data may suffer from some level of *information loss*. Proper care in the aggregation process should be given to significant digits, rounding, averaging, and other arithmetic operations to avoid unnecessary loss of precision.

6. For each cluster (see Section 2.1), there should be an aggregator or set of potential aggregators.

7. Aggregators are (1) executed at a specific time and for a fixed time interval or (2) event-driven.

8. Aggregators may be acquired off the shelf. Note that aggregators may be nonexistent and will need to be homegrown. This may create a problem for huge volumes of data within a NoT.

9. Some NoTs may not have an aggregator, e.g., a single light sensor will send a signal directly to a smart light bulb to turn it off or on.

10. Security is a concern for aggregators (malware or general defects) and for the sensitivity of their aggregated data. Further, aggregators could be attacked, e.g., by denying them the ability to operate/execute or by feeding them bogus data.

11. Reliability is a concern for aggregators (general defects).

2.3.2.1 Actor #1: Cluster (or "Sensor Cluster")

A *cluster* is an abstract grouping of sensors that can appear and disappear instantaneously. Basic properties, assumptions, recommendations, and general statements about cluster include the following:

1. Clusters are abstractions of a set of sensors along with the data they output – clusters may be created in an *ad hoc* manner or organized according to fixed rules.

2. Clusters are not inherently physical.

3. C_i is essentially a *cluster* of the sensor data from $n \geq 1$ sensors $\{d_1, d_2, d_3, \ldots, d_n\}$.

4. C_i may share one or more sensors with C_k, where $i \neq k$, or with other NoTs. This is somewhat important, because competing organizations may be receiving data that they believe to be unique and purposed only for them to receive and not realizing a competitor is also receiving the same sensor data.

5. *Continuous binding* of a sensor to a cluster may result in little ability to mitigate trustworthiness concerns of a real-time NoT if the binding occurs *late*.

6. Clusters are malleable and can change their collection of sensors and their data at any time.

7. The composition of clusters is dependent on what mechanism is employed to aggregate the data, which ultimately impacts the purpose and requirements of a specific NoT.

Note item 4 in the above list; it is subtly important – it relates to business competition for highly valued data.

2.3.2.2   Actor #2: Weight

Weight is the degree to which a particular sensor's data will impact an aggregator's computation. Basic properties, assumptions, recommendations, and general statements about weight include the following:

1. A weight may be hardwired or modified on the fly.

2. A weight may be based on a sensor's perceived trustworthiness, e.g., based on who is the sensor's owner, manufacturer, geographic location where the sensor is operating, sensor age or version, previous failures or partial failures of sensor, sensor tampering, sensor delays in returning data. A weight may also be based on the worth of the data, uniqueness, relation to mission goals, etc.

3. Different NoTs may leverage the same sensor data and recalibrate the weights per the purpose of a specific NoT.

4. It is not implied that an aggregator is necessarily a functionally linear combination of sensor outputs. Weights could be based on other logical insights, such as the following: if sensor A output is greater than 1, use sensor B's output else use sensor C's output.

5. Aggregators may employ AI techniques to modify their clusters and weights on the fly.

6. Weights will affect the degree of information loss during the creation of intermediate data.

7. Redundant sensors may increase a sensor's weight if a grouping of redundant sensor data is in agreement and produces the same result. Repeated sampling of the same sensor might also affect a sensor's weighting, either positively or negatively, depending on the continuity of a particular output value during a fixed time interval.

8. Security concerns for weights are related to possible tampering of the weights.
9. The appropriateness (or correctness) of the weights is crucial for the purpose of a NoT.

A simple aggregator might implement the summation

$$\sum_{i=1}^{x} d_i$$

divided by x,
where
    x is the number of data points, and where the weight for each data point is *uniform*.

2.3.3 PRIMITIVE #3: COMMUNICATION CHANNEL

A *communication channel* is a medium by which data is transmitted (e.g., physical via Universal Serial Bus (USB), wireless, wired, verbal). Basic properties, assumptions, recommendations, and general statements about communication channel include the following:

1. Communication channels move data between computing, sensing, and actuation.
2. Since data is the "blood" of a NoT, communication channels are the "veins" and "arteries", as data moves to and from intermediate events at different snapshots in time. We talk more about the role of snapshots in time in Section 2.3.
3. Communication channels will have a physical or virtual aspect to them, or both. Protocols and associated implementations provide a virtual dimension. Wires provide a physical dimension.
4. Communication channel dataflow may be unidirectional or bidirectional. There are a number of conditions where an aggregator might query more advanced sensors, or potentially recalibrate them in some way (e.g., request more observations per time interval).
5. No standardized communication channel protocol is assumed; a specific NoT may have multiple communication protocols between different entities.
6. Communication channels may be wireless.
7. Communication channels may be an offering (*service* or *product*) from third-party vendors.
8. Communication channel *trustworthiness* may make sensors appear to be failing when actually the communication channel is failing.
9. Communication channels can experience disturbances, delays, and interruptions.
10. *Redundancy* can improve communication channel reliability. There may be more than one distinct communication channel between a computing primitive and a sensing primitive.

11. Performance and availability of communication channels will greatly impact any NoT that has time-to-decision requirements (see the decision trigger primitive in Section 2.5).
12. Security and reliability are concerns for communication channels.

In Figure 2.1, 15 sensors are shown – the blue sensors indicate that these two sensors are 'somehow' failing and at specific times, i.e., they are not satisfying their purpose and expectations. As mentioned earlier, there could be a variety of sensor failure modes, some temporal and some related to data quality. Further, the temporal failure modes for sensors may be actually a result of the transport of that data failing and not the sensors themselves. Consider also that the two failing sensors in Figure 2.1 should probably be assigned lower weights. Figure 2.1 also shows the 15 sensors clustered into three clusters with five unique sensors assigned to each. Figure 2.1 shows the data coming out from each of the three clusters as being inputted to three corresponding aggregators. It is now the responsibility of the three aggregators to turn those 15 sensor inputs into three intermediate data points.

Note the close relationship between clusters and aggregators. For example, in Figure 2.1, aggregator A_1 might be determining how busy restaurant X is. Five independent sensors in C_1 could be taking pictures from inside and outside (parking lot) of X, room temperature measurement in the kitchen, motion detectors from the dining area, sound and volume sensors, light detectors, etc. So while the sensors are certainly not homogeneous, their data is processed to make a new piece of data to address one question with possible results such as is the restaurant busy, not busy, and closed. And aggregators A_2 and A_3 might be doing the same for restaurants Y and Z, respectively. Consider also that an aggregator can be associated with different clusters.

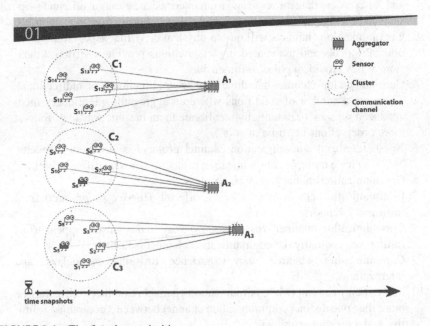

FIGURE 2.1 The first three primitives.

2.3.4 PRIMITIVE #4: εUTILITY (EXTERNAL UTILITY)

An eUtility (external utility) is a software or hardware product or service.[4] The current definition of eUtility is deliberately broad, to allow for unforeseen future services and products that will be incorporated in future types of NoTs yet to be defined. (Breaking this primitive down into subprimitives is future work.) Basic properties, assumptions, recommendations, and general statements about eUtility include the following:

1. eUtilities execute processes or feed data into the overall workflow of a NoT.
2. eUtilities could be acquired off the shelf from third parties.
3. eUtilities may include databases, mobile devices, miscellaneous software or hardware systems, clouds, computers, and CPUs. The eUtility primitive can be subdivided and probably should be decomposed to make this model less abstract.
4. eUtilities, such as clouds, provide computing power that aggregators may not have.
5. A human may be viewed as an eUtility. A human is sometimes considered as a 'thing' in public discourse related to IoT.
6. Data supplied by an eUtility may be weighted.
7. An eUtility may be counterfeit; this is mentioned later in element Device_ ID (see Section 2.3).
8. Nonhuman eUtilities may have Device_IDs; Device_IDs may be crucial for identification and authentication.
9. Security and reliability are concerns for all eUtilities.

Figure 2.2 illustrates using two cloud eUtilities to execute five aggregator implementations. (Note that there is no suggestion here that aggregators must execute on cloud platforms – this was for illustration only – aggregators can execute on any computing platform.) Figure 2.2 shows the addition of one noncloud eUtility, eU_1 (a laptop).

2.3.5 PRIMITIVE #5: DECISION TRIGGER

A decision trigger creates the final result(s) needed to satisfy the purpose, specification, and requirements of a specific NoT. Basic properties, assumptions, recommendations, and general statements about decision trigger include the following:

1. A decision trigger is a conditional expression that triggers an action. A decision trigger's outputs can control actuators and transactions (see Figures 2.3 and 2.4). Decision triggers abstractly define the end purpose of a NoT.
2. A NoT may or may not control an actuator (via a decision trigger).
3. A decision may have a binary output, but there will also be situations where the output of a decision trigger is a nondiscrete output, i.e., a continuum of output values.

---

[4] Standards Developing Organizations (SDOs) have referred to eUtilities as *cyber-entities* [IEEE, p. 2413, "Standard for an Architectural for the Internet of Things (IoT)] and *digital entities* [JTC21 WG10 ISO/IEC 34010, "Internet of Things"] in their draft documents as of this writing.

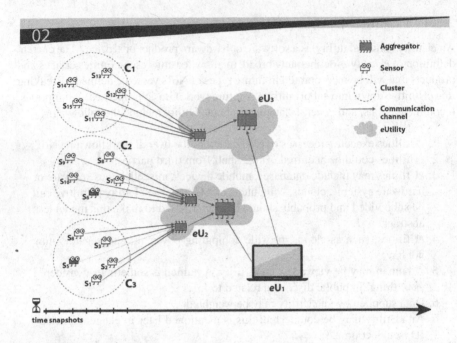

FIGURE 2.2   *e*Utility.

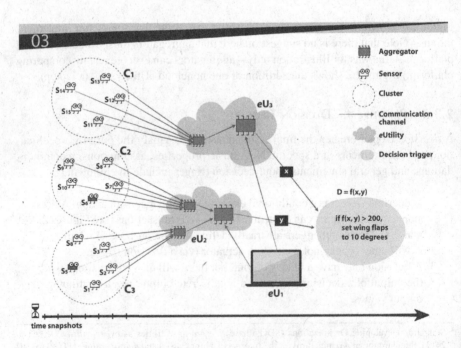

FIGURE 2.3   Decision trigger.

4. A decision trigger may have a built-in adaption capability as the environment element (see Section 2.3) changes.
5. A decision trigger will likely have a corresponding virtual implementation,
   i.e., code.
6. A decision trigger may have a unique owner.
7. Decision triggers may be acquired off the shelf or homegrown.
8. Decision triggers are executed at specific times or may execute continuously
   as new data becomes available.
9. Decision trigger results may be predictions.
10. Analytics could be implemented within decision triggers; however, analytics
    could also be implemented within aggregators (that are executed by eUtilities).
11. If a decision trigger feeds data signals into an actuator, then the actuator
    may be considered as an eUtility if the actuator feeds data back into the
    NoT. Also, an actuator can and often should be considered as a component
    of the environment element (see Section 3). This model treats actuators as
    "consumers" of the outputs from decision triggers. Actuators are 'things',
    but not all things are primitives in this model.
12. A decision trigger may feed its output back into the NoT creating a feedback
    loop[5] (see Figure 2.4).

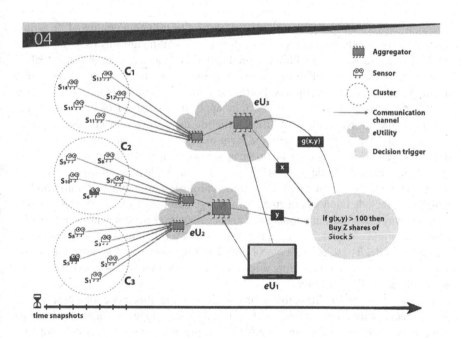

FIGURE 2.4 Decision trigger with feedback.

---

[5] There are other *feedforward* mechanisms in process control theory and implementation that are outside
  the immediate scope of this model.

13. It is fair to view a decision trigger as an *if-then* rule, although they will not all have this form.
14. The workflow up to decision trigger execution may be partially parallelizable.
15. Failure to execute decision triggers at time t_x may occur due to tardy data collection, inhibited sensors or eUtilities, inhibited communication channels, low-performance aggregators, and a variety of other subsystem failure modes.
16. Economics and costs can play a role in the quality of the decision trigger's output.
17. There may be intermediate decision triggers at any point in a NoT's workflow.
18. Decision triggers act similarly to aggregators and can be viewed as a special case of aggregator.
19. Security is a concern for decision triggers (malware or general defects). Other possibilities here might be indirect manipulation of input values to the trigger by tampering with or restricting the input values.
20. Reliability is a concern for decision triggers (general defects). Decision triggers could be inconsistent, self-contradictory, and incomplete. Understanding how bad data propagates to affect decision triggers is paramount. Failure to execute decision triggers at time t_x may have undesired consequences.

Decision triggers are predicates that must be *true* to initiate a command or action. They will frequently be represented as *if-then* rules (although this may not always be the case), and variables within the predicate may represent data or sensor values from parts of the NoT. Reliability is essential, indicating the need for strong testing and assurance of decision triggers. Fortunately, methods and tools for testing access control rules are well developed and may apply to a large set of decision trigger testing problems (Kuhn et al., 2016). Paralleling the structure of decision triggers, access control rules typically take the form of predicates evaluating to two-valued decisions – either *grant* or *deny* in most cases. For example, a building access control system may include a rule such as "*if* subject is an employee *and* time is between 6 am and 9 pm *then* grant access". Access control rules are composed into policies, which may include hundreds of rules with a large number of variables interacting in complex ways. Methods and tools exist to check complex access control policies for consistency and completeness and to generate tests for assurance of policies implemented in code. These access control policy testing tools could be adapted to NoT verification and testing, providing a level of assurance that will be essential for the more critical NoT applications.

Figure 2.4 shows an alternative to any suggestion that this model of a NoT's dataflow is necessarily unidirectional; it depicts a decision trigger that actually feeds its computation for $g(x, y)$ back into the NoT, creating a continuous feedback loop. So, e.g., if new sensor data were fed continuously into a NoT's workflow and dataflow, that data can be combined with the results of previous decision trigger outputs to create updated decision trigger results at later points in time.

2.3.6 ADDITIONAL NOTES ON THE PRIMITIVES

Now, a few additional points concerning the interplay and relationship between the five primitives are as follows. First, sensor feeds aggregator. Second, aggregator executes on various *e*Utilities of a NoT. Third, communication channels are the veins and arteries that connect sensor, aggregator, *e*Utility, and decision trigger with the data that flows between them. And fourth, sensor, aggregator, communication channel, *e*Utility, and decision trigger all have events firing at specific snapshot times; a large challenge for IoT and NoTs is to keep these events in sync, just as for any distributed system.

It is important to treat all events with respect to their temporal location, i.e., their geographical place and at what time. For example, the older the events become, the staler or less interesting they may be relative to immediate decision-making. Historical information may be useful in other calculations such as statistical reporting or model generation.

2.4 THE ELEMENTS

To complete this model, we define six elements: *environment, cost, geographic location, owner, Device_ID,* and *snapshot,* that although are not primitives, are key players in trusting NoTs. These elements play a major role in fostering the degree of trustworthiness[6] that a specific NoT can provide.

1. *Environment*: The universe that all primitives in a specific NoT operate in; this is essentially the *operational profile* of a NoT. The environment is particularly important to the sensor and aggregator primitives since it offers context to them. An analogy is the various weather profiles that an aircraft operates in or a particular factory setting that a NoT operates in. This will likely be difficult to correctly define.
2. *Cost*: The expenses, in terms of time and money, that a specific NoT incurs in terms of the nonmitigated reliability and security risks; additionally, the costs associated with each of the primitive components needed to build and operate a NoT. Cost is an estimation or prediction that can be measured or approximated. Cost drives the design decisions in building a NoT.
3. *Geographic Location*: Physical place where a sensor or *e*Utility operates in, e.g., using RFID to decide where a 'thing' actually resides. Note that the operating location may change over time. Note that a sensor's or *e*Utility's geographic location along with communication channel reliability and data security may affect the dataflow throughout a NoT's workflow in a timely manner. Geographic location determinations may sometimes not be possible. If not possible, the data should be suspect.

---

[6] *Trustworthiness* includes attributes such as security, privacy, reliability, safety, availability, and performance, to name a few.

4. *Owner*: Person or organization that owns a particular sensor, communication channel, aggregator, decision trigger, or *e*Utility. There can be multiple owners for any of these five. Note that owners may have nefarious intentions that affect overall trust. Note further that owners may remain anonymous. Note that there is also a role for an "operator"; for simplicity, we roll up that role into the owner element.

5. *Device_ID*: A unique identifier for a particular sensor, communication channel, aggregator, decision trigger, or *e*Utility. Further, a Device_ID may be the only sensor data transmitted. This will typically originate from the manufacturer of the entity, but it could be modified or forged. This can be accomplished using RFID[7] for physical primitives.

6. *Snapshot*: An instant in time. Basic properties, assumptions, and general statements about snapshot include the following:

   a. Because a NoT is a distributed system, different events, data transfers, and computations occur at different snapshots.

   b. Snapshots may be aligned to a clock synchronized within their own network (NIST, 2015). A global clock may be too burdensome for sensor networks that operate in the wild. Others, however, argue in favor of a global clock (Li & Rus, 2004). This publication does not endorse either scheme at the time of this writing.

   c. Data, without some "agreed upon" time-stamping mechanism, is of limited or reduced value.

   d. NoTs may affect business performance – sensing, communicating, and computing can speed up or slow down a NoT's workflow and therefore affect the "perceived" performance of the environment it operates in or controls.

   e. Snapshots may be tampered with, making it unclear when events actually occurred, not by changing time (which is not possible) but by changing the recorded time at which an event in the workflow is generated or computation is performed, e.g., sticking in a delay() function call.

   f. Malicious latency to induce delays is possible and will affect when decision triggers are able to execute.

   g. Reliability and performance of a NoT may be highly based on (e) and (f).

2.5   ADDITIONAL CONSIDERATIONS

Five additional considerations to the previous material that defined the primitives and elements include:

---

[7] RFID readers that work on the same protocol as the inlay may be distributed at key points throughout a NoT. Readers activate the tag causing it to broadcast radio waves within bandwidths reserved for RFID usage by individual governments internationally. These radio waves transmit identifiers or codes that reference unique information associated with the item to which the RFID inlay is attached, and in this case, the item would be a primitive.

2.5.1 Open, Closed

NoTs can be open, closed, or somewhere in between. For example, an automobile can have hundreds of sensors; numerous Central Processing Unit (CPU)s; databases such as maps, wired communication channels throughout the car, and without any wireless access between any 'thing' in the car to the outside. This illustrates a closed NoT. Such a NoT mitigates nearly all wireless security concerns such as remotely controlling a car; however, there could still be concerns of malware and counterfeit 'things' that could result in reduced safety. (There are issues related to wireless transmission ranges that we ignore here for simplification.) A fully open system would essentially be any 'thing' interoperating with any 'thing', any way, and at any time. This, from a "trustworthiness" standpoint, is impossible to assure since the NoT is unbounded.

Most NoTs will be between these extremes since a continuum will likely exist. The primitives serve as a guidepost as to where reliability and security concerns require additional mitigation, e.g., testing and other technologies related to validation and verification.

2.5.2 Patterns

We envision a future demand for design patterns that allow larger NoTs to be built from smaller NoTs, similar to design patterns in object-oriented systems. In essence, these smaller entities are sub-NoTs. Sub-NoTs could speed up IoT adoption for organizations seeking to develop IoT-based systems by having access to sub-NoT catalogues. Further, the topology of sub-NoTs could impact the security and performance of composite NoTs.

2.5.3 Composition and *Trust*

To understand the inescapable *trust* issues associated with IoT and a specific NoT, consider the attributes of the primitives and elements shown in Table 2.1. The three rightmost columns are our best guess as to whether the pedigree, reliability, or security of an element or primitive creates a trustworthiness risk. Those with question marks (?) are ones do not think we fully understand at this time.

The following table poses questions such as: *what does trust mean for a NoT when its primitives are in continual flux due to natural phenomenon that are in continuous change and while its virtual and physical entities are unknown, partially unknown, or faulty? Or if we have insecure physical systems employing faulty snapshots composed with incorrect assumed environments, where is the trust?*

Such questions demonstrate the difficulty assuring and assessing NoT trustworthiness. We offer the following statement about NoT trustworthiness:

Trust in some NoT A, at some snapshot X, is a function of NoT A's assets ϵ {sensors (s), aggregator(s), communication channel(s), eUtility(s), decision trigger(s)} with respect to the members ϵ {geographic location, owner, environment, cost, Device_IDs, snapshot} when applicable.

TABLE 2.1
Primitive and Element Trust Questions

Primitive or Element	Attribute	Pedigree Risk?	Reliability Risk?	Security Risk?
Sensor	Physical	Y	Y	Y
Aggregator	Virtual	Y	Y	Y
Communication channel	Virtual and/or Physical	Y	Y	Y
eUtility	Virtual or physical	Y	Y	Y
Decision trigger	Virtual	Y	Y	Y
Geographic location	Physical (possibly unknown)	N/A	Y	Y
Owner	Physical (possibly unknown)	?	N/A	?
Environment	Virtual or physical (possibly unknown)	N/A	Y	Y
Cost	Partially known	N/A	?	?
Device_ID	Virtual	Y	Y	Y
Snapshot	Natural phenomenon	N/A	Y	?

2.5.4 NoT Testability

A testability[8] metric that applies here is titled the Domain Range Ratio (Voas & Miller, 1993). This ratio is simply the cardinality of the set of all possible test cases to the cardinality of the set of all possible outputs.

To understand this metric, consider a NoT's decision trigger that forces an actuator to be in one of two states: ('1') or ('0'). (This situation occurs in Figures 2.3 and 2.4.) Further, assume that each output state occurs 50% of the time. Because of this minimal output space size, a fair coin toss also has a 50-50 chance of providing a correct output for any given input, and that's likely less expensive than building a complex NoT. Worse, consider the scenario where '1' and '0' are not evenly distributed, e.g., the specification states that for 1 million unique test cases only 10 should produce a '1' and the other 999,990 should produce a '0'. Here, you could build a NoT to compute this function or you could write a piece of code that simply says: for all inputs output ('0'). This incorrect software implementation is still 99.999% reliable, and you'll almost certainly not discover the code defect with a handful of random tests sampled from the 1 million. In short, testing here has a minimal probability of detecting the faulty logic because each test case has low "detectability" due to the tiny output space and the probability density function for each output.

This shows that because of the decision trigger primitive, specifically purposed NoTs may be inherently untestable without well-placed internal self-tests (assertions[9]) upon the other four primitives to increase testability.

---

[8] *Testability* here refers to the likelihood that defects can be discovered during testing (Voas & Miller, 1995).

[9] Assertions act as "mini-oracles," because they test internal states during execution and add to the argument that the final output was indeed legitimately generated since the internal states that lead to that output were also correct and not dirty.

2.5.5 ENVIRONMENT

As NoTs are likely to rely on factors from outside native environmental boundaries (e.g., sensor readings from a NoT's surrounding external environment, whatever that environment may be), it is equally logical to assume that some form of actuation may be in place to effect or at least influence that external environment as a result of a NoT action. Thus, it is feasible that an *e*Utility may be more than a software-driven virtual entity as is strongly implied but could also be a cyber-physical or purely a mechanical actuator (e.g. control surface hydraulics on an aircraft) that extends an action to influence a NoT environment as opposed to only influencing the internal workings of the NoT itself. This raises the larger question of how a NoT and its environment interact, not only at the element level which is addressed but also, more significantly, at the NoT's operating level.

2.6 RELIABILITY AND SECURITY PRIMITIVE SCENARIOS

The elements lay out key contextual issues related to trustworthiness of a specific NoT. And the primitives are the building blocks of NoTs. Because trustworthiness is such a broad concept, this document has mainly focused on two "ilities" related to the five primitives: security and reliability. People often ask for simple examples of real or hypothetical use cases relating these two "ilities" to each primitive. The following are examples of simple, hypothetical reliability and security scenarios associated with each primitive.

Sensor Reliability: A modern car's speed sensor is exposed to heat, water, and dust (environment). Years later, it starts providing inconsistent readings due to naturally occurring fatigue that induces corrupt sensor data. This is an example of malfunctions caused by environmental conditions.

Sensor Security: A smart building's temperature sensors are easily accessible, and this particular system doesn't provide a means for validating the firmware's authenticity. An attacker substitutes the firmware with one that responds to remote commands. These sensors then become part of a botnet and can contribute to distributed denial-of-service (DDoS) attacks. This is an example of physical tampering and altering firmware.

Aggregator Reliability: In a smart city environment, thousands of sensors transmit data to a series of smart gateways that effectively compress several gigabytes of raw data into meaningful information. A blackout that occurred in part of the city creates an unexpected condition that results in division by zero, which causes the application to keep crashing for the entire duration of the blackout. This is an example of unpredicted conditions that lead to undefined behavior and incorrect output.

Aggregator Security: An attacker introduces a rogue sensor into a network that produces fake readings. These readings are passed as inputs to the aggregator function without any validation. The attacker launches a buffer overflow attack to gain root access to the entire middleware infrastructure (gateway). This is an example of an injection attack or buffer overflow.

Communication Channel Reliability: 'Smart building' sensors for regulating lights and temperature communicate wirelessly via IEEE 802.11 with the rest of the building management system. During a conference, a large number of people are gathered inside a room, having enabled Wi-Fi on their smartphones. Due to overpopulation of the channel, there are frequent disconnections and service degradation. As a result, the sensors are unable to provide readings with their predefined frequency. This is an example of loss of service due to overpopulation and connection problems.

Communication Channel Security: A wearable activity tracker is attached to a person's wrist and measures heart rate and blood pressure. It communicates via Bluetooth Low Energy (BLE) with the wearer's smartphone and forwards the data to a physician. Despite the fact that BLE takes specific actions to randomize the MAC address of the devices, the manufacturer neglected this feature. An attacker with a high-gain antenna can track the presence of the wearer in a crowd and create a movement profile. This is an example of eavesdropping on the communication channel.

eUtility Reliability: A point-of-sale system conducting automatic smart payments depends on a cloud service for verifying the identity of the person using a card. System maintenance of the cloud server happens to occur during business hours, which causes delays in verification. This is an example of system failures that make the resource unavailable and, therefore, service unreliable.

eUtility Security: A 'smart home' has a security camera installed at the front door that sends data to a corresponding cloud application that then forwards notifications and video footage to the homeowner's device after motion at the door is detected. An attacker conducts a DDoS attack on the application provider's servers for 2 hours. They're able to break into the house without the user being notified. This is an example of a DDoS attack.

Decision Trigger Reliability: The logic implemented, e.g., a decision trigger is just a conditional expression, e.g., if a > 100 open garage door. But what if the expression should have been if a > 10 open garage door. This is a reliability problem because the function implemented is incorrect for all values of a between 11 and 99. Decision triggers are likely to be written in code, although this assumption could be relaxed.

Decision Trigger Security: The decision trigger implementation accepts malicious inputs or potentially the outputs from the trigger are sniffed and released to competitors unbeknownst to the legitimate owner of the trigger. Either way, this is an example of data tampering and a loss of data integrity.

Note that there are relatively few standards or best practices for IoT security design and testing, although some related guidance is being developed by the Cyber-Physical Systems Public Working Group (NIST, 2014a) and in documents such as the NIST's *Guidelines for Smart Grid Cybersecurity* (NIST, 2014b).

2.7 SUMMARY

This chapter offers an underlying and foundational science for IoT-based technologies on the realization that IoT involves *sensing, computing, communication,* and *actuation.* We presented a common vocabulary to foster a better understanding of IoT and better communication between those parties discussing IoT. We acknowledge that the Internet is a network of networks, but we believe that focusing on restricted

NoTs in a bounded way gives better traction to addressing *trustworthiness* problems that an unbounded Internet does not. Some may argue that every 'thing' in a NoT is ultimately a service. The primitives definitely offer services; however, because of the combinatorics of mixtures of hardware and software 'things', we prefer to distinguish.

Five primitives and six elements that impact IoT trustworthiness have been presented. Primitives are the building blocks; elements are the less tangible *trust* factors impacting NoTs. Primitives also allow for analytics and formal arguments of IoT use case scenarios. Without an actionable and universally accepted definition for IoT, the model and vocabulary presented here expresses how IoT, in the broad sense, *behaves*.

Use case scenarios employing the primitives afford us quicker recommendations and guidance concerning a NoT's potential trustworthiness. For example, authentication can be used in addressing issues such as geolocation and sensor ownership, but authentication may not be relevant if an adversary owns the sensors and can obtain that information based on proximity. Encryption can protect sensor data transmission integrity and confidentiality including cloud-to-cloud communication, but it might render the IoT sensors unusable due to excessive energy requirements. While fault-tolerant techniques can alleviate reliability concerns associated with inexpensive, replaceable, and defective third-party 'things', they can also be insecure and induce communication overhead and increased attack surfaces. In short, primitives and how they can be composed create a design vocabulary for how to apply existing technologies that support IoT trustworthiness.

These primitives are simply *objects* with *attributes*. The five, along with the context offered by the six elements, form a design *catalog* for those persons and organizations interested in exploring and implementing current and future IoT-based technology.

2.8 ADDITIONAL TAKEAWAY MESSAGES

1. 'Things' may be all software, hardware, a combination of both, and human.
2. A NoT may or may not employ 'things' connected to the Internet.
3. The number of 'things' in a NoT fuels functional complexity and diminishes *testability* unless *observability* is boosted by internal test instrumentation (assertions).
4. NoTs bound scalability and complexity and, therefore, enhance arguments for trustworthiness since assurance techniques generally offer better efficacy to less complex systems.
5. Known threats from previous genres of complex software-centric systems apply to NoTs.
6. Security flaws and threats in NoTs may be exacerbated by the composition of third-party 'things'. This creates an *emergent* class of security 'unknowns'.
7. NoTs may have the ability to self-organize, self-modify, and self-repair when AI technologies are introduced, e.g., neural networks, genetic algorithms, and machine learning. If true, NoTs could potentially rewire their security policy mechanisms and implementations or disengage them altogether.

8. "After the fact" forensics for millions of composed, heterogeneous 'things' are almost certainly not possible in linear time.

9. 'Things' will be heterogeneous. Counterfeiting of 'things' may lead to seemingly nondeterministic behavior making testing's results appear chaotic. Counterfeit 'things' may lead to illegitimate NoTs.

10. Properly *authenticating* sensors may be a data integrity risk, e.g., the 'who is who' question. 'Things' may deliberately misidentify themselves.

11. 'Things' may be granted a nefarious and stealth connection capability, i.e., coming and going in instantaneous time snapshots, leaving zero *traceability*. This is a "drop and run" mode for pushing external data into a NoT's workflow. This may be mitigatable via authentication, cryptography, and possibly others. "Drop and run" affects trustworthiness.

12. *Actuators* are 'things'. If they are fed malicious data from other 'things', issues with life-threatening consequences are possible if the actuator operates in a safety-critical environment.

13. NoTs have workflows and dataflows that are highly *time*-sensitive – NoTs need communication and computation *synchronization*. Defective local/global clocks (timing failures) lead to deadlock, race conditions, and other classes of system-wide, NoT failures.

2.9　ACRONYM GLOSSARY

AC	Alternating current
AI	Artificial intelligence
BLE	Bluetooth Low Energy
CPU	Central Processing Unit
DDoS	Distributed denial-of-service attack
FPGA	Field-programmable gate array
IEC	International Electrotechnical Commission
IEEE	Institute of Electronics and Electrical Engineers
IoT	Internet of Things
ISO	International Standards Organization
MAC	Media access control
NIST	National Institute of Standards and Technology
NoT	Network of Things
PHM	Prognostics and Health Management
RFID	Radio Frequency Identification
SDO	Standards Developing Organization
USB	Universal Serial Bus
Wi-Fi	Any wireless local area network product based on IEEE 802.11

FURTHER READING

Kuhn, D. R., Hu, V., Ferraiolo, D. F., Kacker, R. N., Lei, Y., "Pseudo-exhaustive testing of attribute based access control rules," *International Workshop on Combinatorial Testing at the 2016 IEEE Ninth International Conference on Software Testing, Verification and Validation Workshops (ICSTW)*, Chicago, Illinois, April 10–15, 2016.

Li, Q., Rus, D., "Global clock synchronization in sensor networks," *Twenty-Third Annual Joint Conference of the IEEE Computer and Communications Societies (INFOCOM 2004)*, Hong Kong, March 7–11, 2004, pp. 564–574. doi:10.1109/INFCOM.2004.1354528.

NIST, *Cyber-Physical Systems Public Working Group Workshop*, [Web page], National Institute of Standards and Technology, August 11–12, 2014a. Accessed 13 July 2016. http://www.nist.gov/cps/cps-pwg-workshop.cfm.

NIST, *The Smart Grid Interoperability Panel—Smart Grid Cybersecurity Committee, Guidelines for Smart Grid Cybersecurity*, NIST Interagency Report (NISTIR) 7628 Revision 1. National Institute of Standards and Technology, Gaithersburg, MD, September 2014b, 668 pp. doi:10.6018/NIST.IR.7628r1.

Voas, J. M., "Networks of 'Things'," NIST Special Publication 800-183, 2016. https://csrc.nist.gov/publications/detail/sp/800-183/final.

Voas, J. M., Miller, K. W., "Semantic metrics for software testability," *Journal of Systems and Software*, vol. 20, no. 3, March 1993, pp. 207–216. doi:10.1016/0164-1212(93)90064-5.

Voas, J. M., Miller, K. W., "Software testability: the new verification," *IEEE Software*, vol. 12, no. 3, May 1995, pp. 17–28. doi:10.1109/52.382180.

3 Smart Cities[1]

3.1 INTRODUCTION

The United Nations estimates that almost 70% of the world population will live in urban areas by 2050 (United Nations, 2018). Consequently, the urban environments, which are already crowded and chaotic, should be further improved and prepared to accommodate that higher number of citizens sustainably and feasibly.

Cities will then be required to become smarter, i.e., to be transformed and connected via their physical, social, and business infrastructures using Information and Communication Technology (ICT) to leverage the collective services to the society (Harrison et al., 2010; Mohanty, Choppali & Kougianos, 2016). Smarter means that the services should intensely rely on technology infrastructure, particularly software. That infrastructure is composed of a diversity of individual systems and physical objects (e.g., sensors, actuators, and tags). For instance, in a smart traffic system, sensors can detect the presence of a person in a crosswalk. The traffic lights turn to red and the autonomous cars stop movement through smart break systems (actuators). People cross the street approximating to tags to notify the traffic should still be interrupted. As soon as no presence is detected, the traffic lights turn green and the car's actuators restart their movement. Those systems and objects interoperate to provide solutions for the population, such as

i. offering transportation options that could be more economical or optimized about time and lower risks, as it has been implemented in Singapore (Asiag, 2020);
ii. interoperating systems to provide a collective sense of safety, avoiding collisions between cars, between these and public buildings or between cars and pedestrians, as in the application proposed by Taylor, Siebold & Nowzari (2020);
iii. exchanging and sharing data/information or energy to prevent electricity, water, or other natural resources waste, such as sensors measuring the energy wasting in a neighborhood and alerting users they should save power; or
iv. offering solutions for the population using artificial intelligence (AI) algorithms that could deliver more optimized results than using ad hoc strategies, such as smart solutions to find faster or more economical routes to achieve a destination.

Hence, turning smart cities into reality inherently leads to the intensive use of technology in all the public sectors.

[1] This chapter was contributed by Valdemar Vicente Graciano Neto and Renato F. Bulcão-Neto.

DOI: 10.1201/9781003027799-3

As a matter of consensus, the Internet of Things (IoT) is the technical backbone intended to underpin smart cities (Alampalli & Pardo, 2014). Under that perspective, a smart city comprises a network of interconnected physical objects so-called "things", i.e., various physical components that embed some portion of the software, including electronics, sensors, actuators, a microcontroller, or a microprocessor-based device, embedded computing devices, and Radio Frequency Identification (RFID)-based devices. Things can transmit and receive data over the smart city network infrastructure to collect information from the environment, control other devices, exchange and process data, or interact with a user.

Given that first glance, we can deepen the IoT theoretical basis that supports smart cities and exploit the possible synergies that can emerge from combining those topics to establish the forthcoming evolution stage of what we know as cities.

This chapter thus involves offering a comprehensive background on smart cities besides explaining how IoT is a critical topic in that domain. In the next section, we provide a motivational example that will guide our discussion for achieving that goal.

3.2   MOTIVATIONAL EXAMPLE

Our motivational example comprises a smart home. Smart homes (or smart houses) are important components of a smart city and materialize an ideal scenario to explain how IoT and smart systems, in general, are tightly intertwined. In our example, the smart home is owned by an older adult, which demands a sort of specific services. That house consists of a domestic environment monitored by several sensors to provide elderly people with a fall detection and notification service. Falls are a really important topic, being one of the main reasons that lead to injuries and deaths of older adults (Burns, Stevens & Lee, 2016). As such, that smart home environment has been equipped with IoT technologies to monitor and detect falls and notify smart emergency services accessible by citizens in a city (Chaudhuri, Thompson & Demiris, 2014; Gutierrez-Madroñal et al., 2019).

In our scenario, an elderly person can be monitored by wearable and nonwearable devices. Wearable devices can be used in their everyday life, such as smartphones and smartwatches (Mauldin et al., 2018). Conversely, when the person is in the bathroom, nonwearable devices such as conventional cameras, depth cameras (e.g., Kinect), and Doppler radar can monitor her/his actions to instantaneously detect an eventual fall (Anishchenko, Zhuravlev & Chizh, 2019). Although a small percentage of falls occur in the bathroom, these are at least twice as likely to result in an injury as falls in other in-home places (Mulley, 2001; Stevens, Mahoney & Ehrenreich, 2014). Moreover, most of those falls are not detected, once people frequently take off the wearable devices for a shower, for example.

Regardless of where it happens, once a fall is detected, the smart home environment should notify emergency services so that health professionals could assist the elderly, sending an ambulance, and removing her/him as soon as possible. The smart home system then notifies an emergency service. The emergency service sends an ambulance that communicates with the smart traffic monitoring system to find the shortest and fastest path until the elderly's home. Meanwhile, the traffic monitoring system monitors the ambulance's route in response to likely traffic problems. Once the ambulance reaches the destination, the older adult is then removed.

The ambulance communicates with the smart public health system to find the closest and most equipped hospital that could receive the elderly for emergency treatment and exams. The destination is a smart hospital that alerts the elderly's emergency contacts as soon as she/he is admitted. Meanwhile, a wireless body sensor network continuously monitors the patient's vital signs to notify in-hospital health professionals whenever her/his health conditions worsen.

Using this example, in the following sections, we will bring the theoretical and technological basis that converges IoT and smart cities, the anatomy of a smart city under IoT perspective, and discuss the challenges that remain on the adoption of IoT for smart cities.

3.3   BACKGROUND

A smart city is a complex network of several independent and interoperating systems. Smart cities use ICTs, particularly IoT, and other means to improve life quality and efficiency of urban operations and services while ensuring that it meets the needs of present and future generations with respect to economic, social, and environmental (sustainability) aspects (ITU-T, 2014).

Standards for smart cities and their different components, including smart grids, the IoT, eHealth, and intelligent transportation systems (ITS), have been developed. A specific example of such a standard is ISO 37120 (ISO, 2014), which defines 100 city performance indicators, 46 core and 54 supporting, some of which include economy, education, energy, and environment. IoT can play an important role in smart cities by enabling the collection of some of those indicators. For instance, sensors could collect environmental or energy data, such as humidity and air quality, or the energy expenditure in a neighborhood. Components of a smart city include smart infrastructure, smart buildings, smart transportation, smart energy, smart healthcare, smart technology, smart governance, smart education, and smart citizens (Mohanty, Choppali & Kougianos, 2016; Manzano, Graciano Neto & Nakagawa, 2020).

A smart city is actually formed by a high diversity of smart components and systems. Smart houses, smart traffic control systems, and autonomous cars are other examples of systems that are part of the smart city systems. They are networked together and need to interoperate at a certain level to make the city work. By "certain level" we mean that not all the systems directly interoperate. For instance, smart buildings and smart traffic control systems do not have many reasons to interoperate, while, in parallel, a traffic control system and autonomous cars certainly should be able to converse.

As shown in Figure 3.1, a smart city is structured on many levels. In general, these levels are materialized by geographic coverage. At the bottom level, we have isolated systems themselves. These systems are majorly private properties, such as smart houses, smart buildings, autonomous cars, and smart hospitals. In the next level, we find the first level of interoperability that, in general, directly links the private systems to other external systems, such as the smart home from our motivational example that is linked to a public emergency service. Actually, multiple systems within a smart home environment can be linked to different systems externally, such as wearable devices linked to emergency services and electrical devices linked to

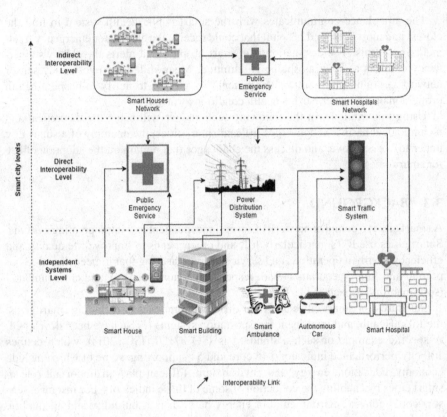

FIGURE 3.1   The concept of a smart city.

a home power distribution management system, which is part of a public or private smart grid. If we consider that each of these smart environments can be considered an independent system that interoperates to other external and independent systems, it is easier to understand why smart cities are frequently called systems of systems (Cavalcante et al., 2017), i.e., a collection of multiple independent systems that interoperate to achieve common established goals. However, since this is not the focus of this chapter, we only provide you with supplementary references for further reading (Cavalcante et al., 2017; Graciano Neto et al. 2018, 2020).

Expanding this notion, we reach the indirect interoperability level, in which several systems start to be connected to form the smart city architecture, such as the entire network formed by the smart homes connected to the smart hospitals via the emergency service. Hence, we observe the form of the city being outlined due to the multiple interoperability links established among the constituent systems. And how is the relationship established between IoT and smart cities? We deepen this discussion in the next section. For now, we recall the important IoT concepts that will be needed for this discussion.

Smart cities rely on IoT for the establishment of the offered services. Smart cities are actually an instance of a recurrent domain where IoT is applied and share some important dimensions with IoT, namely: (1) complexity, (2) everything in IoT (and in a smart city) is a service, (3) spatial dimension, (4) temporal dimension, (5) intelligence and Big Data, and (6) architecture (Perera et al., 2014, 2018).

Complex systems are those with many independent and interdependent components, which interact in nonlinear ways (behavior cannot be expressed as the direct conjunction of the activity of individual components) and have interdependencies that are difficult to describe, predict, and design (Alampalli & Pardo, 2014). The IoT includes many physical objects (e.g., sensors and actuators) that interact autonomously and highly dependent on their capabilities, such as storage, processing, and memory (Khanna and Kaur, 2020). Both smart cities and IoT-based systems are composed of several communicating systems that form large networks that can be hard to manage, which characterizes a complex system.

Second, **everything is a service**. IoT may demand a tremendous amount of infrastructure (e.g., sensing resources) to ease a smart city's deployment (Sheng et al., 2013). Highly interrelated with IoT, the cloud computing paradigm has already shown that consuming resources as a service (platform, infrastructure, or software) is highly efficient, scalable, and easier to use. Thus, this is actually an emerging but already recurrent paradigm that structures everything as a service, with functionalities (e.g., sensing and analytics) that are offered to be invoked by other systems and applications via well-defined interfaces (Tanna, Kumar & Karthika, 2017). This can be clearly observed in Figure 3.1, where the emergency service is available via an interoperability link between the elderly's smart home system and the emergency service itself.

Third, both systems are **spatially structured**, i.e., they have geographical coverage and can be spread along a region, forming alliances, such as the network formed by the smart hospital, the emergency service, and the ambulance.

Regarding the **temporal dimension**, the IoT infrastructure in a smart city may manage many parallel and concurrent events due to the vast number of interactions. In smart cities, real-time data processing is crucial when a system must be notified immediately after a significant event occurs, such as the existing interaction between the smart home environment and the emergency service. In brief, both space and time rule the different types of events taking place in every IoT use case, such as smart cities (Arruda & Bulcão-Neto, 2019).

Moreover, both IoT and smart city systems should provide **intelligent services**, i.e., could use AI applications to process the large volume of data that forms a real Big Data structure. In that sense, Big Data analytics contributes to making the smart city's constituent systems useful and beneficial to its citizens. In our scenario, traffic sensors spread around the city continuously feed the traffic control system that analyzes such data to guide the ambulance through the safest paths until the hospital.

Finally, for both, an **architecture** emerges, i.e., a composition of physical objects with their software counterparts that, together with their connections, assume a structure that can be analyzed and monitored in a secure and scalable fashion.

For establishing the relation between IoT and smart cities, we need to have two main ideas in mind: the key functional features addressed in an IoT architecture and the anatomy of a smart city application (Weyrich & Eber, 2016; Lynn et al., 2020). Since the latter we will discuss in the next section, you only need to understand that the things comprise the sensing (and acting) interface by which any smart system that is part of a smart city senses (and/or acts over) its surrounding environment to collect data and transmit to the other software layers. About the former, interoperability, scalability, security and privacy, data management, among others, constitute fundamental functionalities of an IoT reference architecture addresses that, in turn, guides the development of integrated and multitiered IoT systems, such as smart cities (IEEE, 2019a). The next section discusses the anatomy of this structure.

3.4 THE ANATOMY OF A SMART CITY UNDER IoT PERSPECTIVE

As we have discussed, in a smart city, all the systems are networked together to form a real "internet of things" directly or indirectly. These things appear in different granularities and dimensions. Hence, we deepen on the smart home within the smart city example to discuss how IoT and smart cities are intertwined for the sake of exemplification.

Let us consider a **general data life cycle** in IoT applications: data acquisition, modeling, reasoning, and dissemination (Perera et al., 2014; Maranhão & Bulcão-Neto, 2016). Data is acquired from physical or virtual sensors (acquisition) and processed and stored in an internal and meaningful representation (modeling), such as attribute-value pairs. Data preprocessing techniques over modeled data may precede inference processes, which apply decision models over these data (reasoning). Then, actuators perform the decided actions or inference results trigger event notifications to a third-party system (dissemination).

Noteworthy that depending on the networking, processing, and storage capabilities of a given component, these workflow steps may occur at multiple and different components of an IoT system (Lynn et al., 2020): at the cloud (centralized), at a device, or in an intermediate layer between the devices and the cloud (the fog).

Cloud computing is central to most IoT systems, like smart cities. It provides sensor data collection, storage, access, analytics, and actuation support services, besides administrative functions, such as device and user account management and reporting capabilities. Fog computing, in turn, has been recently a commonplace computing paradigm in IoT. It is a complement to cloud computing, consisting of fog nodes (e.g., devices) that offer cloud services closer to the users (Giannoutakis et al., 2020). Hence, fog computing reduces communication latency and data transfer costs to a cloud, enabling real-time applications and services.

Figure 3.2 illustrates a general IoT-based architecture for smart cities, mostly influenced by the ITU-T's reference model for the IoTs (ITU-T, 2012). The perception layer comprises sensors, actuators, tags, and general devices (e.g., everyday appliances), directly or indirectly (e.g., through a gateway) associated with the network layer. The network layer implements networking and transport capabilities that enable devices to transmit information securely and with low latency, such as authentication, authorization and accounting, mobility management, and transport of application-specific data. This layer is the core of an IoT system because it may connect smart things, network devices, and servers through multiple networks, such as the Internet, sensor networks, and communications networks. Despite its importance, details about the network layer are out of the scope of this chapter.

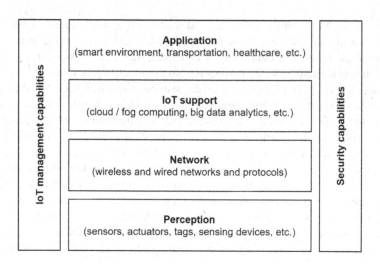

FIGURE 3.2 Internet of Things (IoT)-oriented architecture for smart cities.

On top of this, the primary IoT support layer's role is to provide generic functions to diverse applications in a smart city scenario. These functions may include cloud and fog computing services, such as data processing, data storage, networking, and data analytics (Santos, Monteiro & Endo, 2020). Due to the differences in handling large volumes of different forms of data with velocity and veracity, the design decision by cloud or fog depends on the application's latency requirement. Fog computing is more suitable for low latency applications (Shrivastava & Pandey, 2020), such as the in-home patient health monitoring situation described in Section 9.2. Besides generic support capabilities, the IoT support layer may also cater to specific requirements of varied applications.

Finally, the application layer provides the user interface of smart cities vertical applications with rich visualizations of analysis results. Examples of applications include, but are not limited to, smart government, smart environment, smart transportation, smart utilities, and smart services (e.g., healthcare; Cui et al., 2018).

IoT management capabilities include generic resources for IoT use cases, such as device management, local network topology management, traffic and congestion management, and resource reservation for critical data. Besides, IoT applications' specific requirements may also be supported. Observe that IoT applications' generic and specific security demands are also considered, e.g., privacy protection at the application layer and authorization, authentication, data confidentiality, and integrity protection at the application, network, and device layers.

The wearable and nonwearable devices at home, the traffic lights spread around the city, the ambulance's GPS, and the patient's body sensor network reside in the perception layer.

Due to its sensing, networking, and processing capabilities, older adults' smartphone may be adequate to run a fall monitoring service. However, the same does not apply to the Doppler radar-based device, mainly regarding its processing power. The smartphone may acquire the elderly's address, emergency contacts, and movement data (including the radar device's data) and send it to the cloud. Data may be preprocessed, stored, and analyzed through a specific fall detection service in the cloud. Whenever the movement data analysis concludes that a fall has occurred, the fall detection service automatically notifies the smart city emergency service.

Observe that alternative designs could also be planned, with the fall monitoring service implemented and deployed at the elderly's home. Attention should be paid to the latency needed to conduct the acquired data until the processing center in the cloud (performance). Moreover, problems related to safety and quality of service may arise since a missing connection could prevent the data from being processed, preventing the emergency service to be called, and the domestic accident could become fatal. Hence, architectural alternatives should be exploited so that the resulting system meets each type of stakeholder's requirements and needs. For instance, a fog node serving multiple houses in a city region might acquire, model, and preprocess elderlies' georeferenced movement data. Then, it might automatically send this preprocessed data to a fall detection service resident in the cloud.

Back to the smart city emergency service notification, it sends an ambulance to the house where a fall took place. Throughout its route, the ambulance is monitored by smart traffic lights spread around the city, measuring and analyzing traffic data (e.g., drivers' speed) as it is collected. As any traffic accident notification should be on time for the ambulance's emergency staff, the traffic lights themselves may make decisions to reorganize routes whenever it is necessary. The collected data can be sent to the cloud later on for more long-term and in-depth analyses, e.g., for traffic planning purposes.

The smart public health system runs in the cloud. It collects and analyzes information from other healthcare institutions in the city to decide to which hospital the emergency team must drive. Once again, smart traffic lights help the ambulance to reach the hospital.

Finally, the patient's emergency contacts are communicated by a private service running at the hospital infrastructure. During the hospitalization period, the patient's vital signs are sensed by a wireless body sensor network that sends them to a vital sign analyzer service running at the hospital's private cloud. Whenever the analysis of vital signs indicates the patient's health condition degradation, that service notifies health professionals' smartphones in charge.

From that discussion and merging Figures 3.1 and 3.2, we can observe that IoT and smart cities converge that, in a domestic environment or the public spaces, the IoT sensing layer is the frontier that connects the smart city system to the public (or domestic) environment. In parallel, modeling and reasoning layers are the IoT components that support the data representation, the services implementation, and the interoperability between the systems that compose a smart city. From a layered architectural perspective, it is evident that an ideal smart city system should understand the capability differences and perform data management accordingly with effectiveness and efficiency.

Since we acquired a big picture of how IoT and smart cities have intersections, we can now discuss the challenges for combining technologies to become smart cities a widespread reality. The following section brings such discussion.

3.5 CHALLENGES FOR USING IoT AS THE BACKBONE OF SMART CITIES

Several challenges remain to bring smart cities to reality. Some of them are discussed as follows.

Interoperability: Interoperability per se is a grand challenge for several disciplines, ranging from computer networks, passing by information systems, and reaching software engineering (Boscarioli, Araujo & Maciel, 2017; Maciel et al., 2017). For smart cities and IoT, this is not different. Recalling the basic notion, interoperability consists of the ability of two or more systems to exchange data/information and use the data/information exchanged. Then, interoperability is a challenge because interoperability links should enable effective communication between technology elements, and those elements can be highly heterogeneous. The interoperating systems or components maybe were codified in different programming language paradigms, or are deployed in different hardware architectures or operating systems, or the network technologies to communicate are different and not compatible (examples), or the difference can even be at business level, with a person being represented with a different set of attributes in two different systems. In those cases, *how to establish communication between these systems and enable them to exchange data and use the data exchanged effectively?*

In IoT-based systems (as smart cities), these challenges can be even maximized once: (1) the metrics used to sense the environment can be different among things and demand conversions (e.g., Fahrenheit or Celsius in different devices), (2) the data representation can be different, which could cause redundancy or inconsistencies in the data being represented in the data modeling step (e.g., ontologies and object-oriented models), and (3) the reasoning step demands a sort of intelligence to deal with a potential high diversity of types of data and their associated representation (e.g., rule- and ontology-based reasoning).

Solutions have been proposed over the years, such as realizing the full interoperability (Maciel et al., 2017) and the automatic synthesis of domain-specific middleware (Costa et al., 2017). In the former, supporting all interoperability levels (syntactic, semantic, pragmatic, dynamic, and organizational or conceptual) for specific domain applications could be considered as full interoperability. In the latter, the idea comprises the automatic synthesis of middleware, i.e., a software component capable of abstracting the heterogeneities mentioned above, for specific domains. For instance, in a smart home environment, a software component could analyze the surrounding things and automatically synthesize a component capable of collecting the data representation formats, such as automatically establishing mappings that could support a communication without inconsistencies between the involved components. This could avoid inconsistencies in fall detection, for instance. However, both topics are still a matter of investigation but great promises to solve the problem for IoT and smart cities.

A Quality Model for IoT and Smart Cities: We mentioned the possible consequences of different architectural designs over the quality exhibited by a smart home system during this chapter. The complexity of trading-off the multiple quality attributes involved in a city-scale system can be even more challenging. Indeed, a quality model for smart cities and the consequences over the IoT-based systems that compose it is still lacking and can bring significant results in the forthcoming years. Particularly, the quadruple "protection, security, privacy, and safety" (IEEE, 2019a; Spinellis, 2017); availability (Delécolle et al., 2020); responsiveness (Motta, de Oliveira & Travassos, 2019); and other attributes can be further explored to assure the quality of the services provided by smart cities' users.

Regarding privacy, in particular, a smart city collects and processes personal, enterprise, and governmental data 24/7 that an unauthorized person should not be able to access. Besides cryptography, anonymization, and access control, two factors are determinants for a comprehensive privacy protection solution: time and space (Arruda & Bulcão-Neto 2019; Perera et al., 2020). Access to the elderly's data is immediately granted to an emergency service when a fall occurs and remains until her/his admission at the hospital. From this moment, the patient's location is notified to emergency contacts, whereas her/his health conditions are available only by health professionals in charge during the hospitalization period. Finally, the closer an IoT device is to the end user (e.g., the patient's smartphone), the more challenging it is to protect the privacy of consumers before their sensitive data "leaves" home (Cui et al., 2018).

A Standard Reference Architecture for Smart Cities: The IEEE's IoT architecture working group recently developed an architectural framework for the IoT called "P2413", motivated by concerns commonly shared by systems stakeholders across multiple IoT domains (IEEE, 2019a). In general, these concerns relate to disjointed, domain-specific, and often redundant standardization efforts. The P2413 incorporates a collection of architecture viewpoints elaborated to form the framework description body: a reference model, a reference architecture, and a blueprint for data abstraction.

An active project of the same IEEE working group leveraged on the IEEE P2413 is the standardization of a Reference Architecture for Smart Cities (RASC; IEEE, 2019b). RASC's goals include defining different vertical applications in the smart city (e.g., water and waste management, environment monitoring, smart buildings, and eHealth), commonalities between these vertical applications, and an intelligent and integrated operations center in a smart city use case. Besides, RASC also proposes a four-tiered architecture for smart cities (device, communication network, IoT platform, and application) and defines the relationships with attributes specific to cloud computing technologies and Big Data analysis. By this writing, RASC is still an IEEE draft standard, thus under regular revisions.

Software-Engineering Smart Cities: It is common sense that a considerable investment in networking infrastructure is needed to make a smart city a reality. Nevertheless, networking is not enough (Weyrich & Ebert, 2016). Software is massively pervasive in smart cities, covering data acquisition, communication, representation, storage, reasoning, distribution, retrieval, and visualization, among others. Furthermore, the multidisciplinary nature of IoT and the multitude of stakeholders in a smart city challenge current software engineering practices (Santana et al., 2018; Motta, de Oliveira & Travassos, 2019).

In a smart city use case, requirements should be elicited from citizens and system stakeholders using a more comprehensive approach with business, usage, functional, and implementation viewpoints (Fraile et al., 2019). A business viewpoint conforms to a value-driven model with quantitative performance indicators, such as business value, expected return of investments, and maintenance costs. The usage perspective identifies parties, actors, roles, tasks, and workflows involved in different smart city use cases. The functional viewpoint's output is the decomposition of each constituent system into functional parts, describing its structure and interrelations, interfaces, and interactions between its internal components and other constituent systems. This viewpoint should also include data model definitions, security requirements, and the alignment of tasks to functional components and implementation components. Finally, the implementation viewpoint should provide details on how an asset will be deployed to the computing infrastructure. All of these multiple points of view shall be very well coordinated in a smart city development project. However, in such a large-scale and complex system-of-systems scenario, there is still a road to paving on representing, describing, and integrating systems in the light of software requirements (Motta, de Oliveira & Travassos, 2019). As pointed out by Spinellis (2017), the design and construction of integrated IoT systems, such as smart cities, require complex adaptation layers due to the requirements' dynamic nature as these systems operate. The scale of a smart city also poses challenges to software verification and validation techniques, including simulation (Larrucea et al., 2017; Graciano Neto et al., 2018). Finally, as IoT technologies and real-world smart cities' experiences accumulate, more research on software engineering will be needed, including general models and methodologies and associated tool support (Morin, Harrand & Fleurey, 2017; Santana et al., 2018; Zambonelli, 2017; Motta, de Oliveira & Travassos, 2019).

3.6 CONCLUSION

In this chapter, we discussed the intersection between the IoT and smart cities. We introduced the general structure of a smart city and recalled how an IoT application is composed to show how these two emerging topics are intimately related. We used a motivational example of a smart home environment with things to detect falls so that elderly people could be automatically assisted with an emergency service. From that example, we showed that the IoT sensing layer is the frontier that connects the smart city system to the public (or domestic) environment, drew our discussion, and showed, at the end, the open research avenues that are still available to be explored.

Smart cities and IoT are closely related. The latter is actually considered the backbone of the former, which makes them share several characteristics and also challenges. As long as IoT technologies and its associated challenges, such as a full interoperability, are solved, smart cities will instantaneously be benefited from those advances and be refined to offer more elaborated services to the society.

FURTHER READING

Anishchenko, L., Zhuravlev, A., Chizh, M., Fall detection using multiple bioradars and convolutional neural networks. *Sensors*, vol.19(24), 2019, p. 5569.

Alampalli, S., Pardo, T., "A study of complex systems developed through public private partnerships," *Proceedings of the 8th International Conference on Theory and Practice of Electronic Governance – ICEGOV '14*, 2014, pp. 442–445. doi:10.1145/2691195.2691212.

Arruda, M. F., Bulcão-Neto, R. F., Toward a lightweight ontology for privacy protection in IoT. *Proceedings of the 34th ACM/SIGAPP Symposium on Applied Computing (SAC '19)*. *Association for Computing Machinery*, New York, NY, 2019, pp. 880–888. doi:10.1145/3297280.3297367.

Asiag, J. J., "8 Smart cities lead the way in advanced intelligent transportation systems," Otonomo, 2020. Available at: https://otonomo.io/blog/smart-cities-intelligent-transportation-systems/

Boscarioli, C., Araujo, R. M., Maciel, R. S. P., *I GranDSI-BR – Grand Research Challenges in Information Systems in Brazil 2016–2026*. Special Committee on Information Systems (CE-SI). Brazilian Computer Society (SBC), Porto Alegre, 2017, 184 p. ISBN: [978-85-7669-384-0].

Burns, R., Stevens, J. A., Lee, R., "The direct costs of fatal and non-fatal falls among older adults – United States," *Journal of Safety Research*, vol. 58, 2016, pp. 99–103.

Cavalcante, E., Cacho, N., Lopes, F., Batista, T., "Challenges to the development of smart city systems: a system-of-systems view," *Proceedings of the Brazilian Symposium on Software Engineering*, Fortaleza, 2017, pp. 244–249.

Chaudhuri, S., Thompson, H., Demiris, G., "Fall detection devices and their use with older adults: a systematic review," *Journal of Geriatric Physical Therapy*, vol. 920(37), 2014, p. 178.

Costa, F. M., Morris, K. A., Kon, F., Clarke, P. J., "Model-driven domain-specific middleware," *Proceedings of the International Conference on Distributed Computing Systems*, vol. 2017, Atlanta, GA, 2017, pp. 1961–1971.

Cui, L., Xie, G., Qu, Y., Gao, L., Yang, Y., "Security and privacy in smart cities: challenges and opportunities," *IEEE Access*, vol. 6, 2018, pp. 46134–46145. doi:10.1109/ACCESS.2018.2853985.

Delécolle, A., Lima, R. S., Graciano Neto, V. V., Buisson, J., "Architectural strategy to enhance the availability quality attribute in system-of-systems architectures: a case study," *Proceedings of the System of Systems Engineering Conference 2020*, Budapest, Hungary, 2020, pp. 93–98.

Fraile F, Sanchis R, Poler R, Ortiz A. Reference models for digital manufacturing platforms. *Applied Sciences*. vol.9(20), 2019, p. 4433.

Giannoutakis, K. M., Spanopoulos-Karalexidis, M., Filelis Papadopoulos, C. K., Tzovaras D., "Next generation cloud architectures." In: Lynn T., Mooney J., Lee B., Endo P. (eds) *The Cloud-to-Thing Continuum*. Palgrave Studies in Digital Business & Enabling Technologies. Palgrave Macmillan, Cham, 2020. doi:10.1007/978-3-030-41110-7_2.

Graciano Neto, V. V., Manzano, W., Kassab, M., Nakagawa, E. Y., "Model-based engineering & simulation of software-intensive systems-of-systems: experience report and lessons learned," *Proceedings of the European Conference on Software Architecture (Companion)*, vol. 27, 2018, 1–27.

Graciano Neto, V. V., Santos, R. P., Viana, D., Araujo, R., "Towards a conceptual model to understand software ecosystems emerging from systems-of-information systems." In: Santos R., Maciel C., Viterbo J. (eds) *Software Ecosystems, Sustainability and Human Values in the Social Web*. WAIHCWS 2017, WAIHCWS 2018. Communications in Computer and Information Science, vol. 1081. Springer, Cham, 2020. doi:10.1007/978-3-030-46130-0_1.

Gutierrez-Madroñal, L., Blunda, L. L., Wagner, M., Medina-Bulo, I., "Test event generation for a fall-detection IoT system," *IEEE Internet of Things Journal*, vol. 6, 2019, pp. 6642–6651.

Harrison, C., Eckman, B., Hamilton, R., Hartswick, P., Kalagnanam, J., Paraszczak, J., Williams, P., "Foundations for smarter cities," *IBM Journal of Research and Development*, vol. 54, no. 4, 2010, pp. 1–16.

IEEE P2413, "Standard for an architectural framework for the Internet of Things (IoT)," IEEE Standards Association, 21 May 2019a. Accessed November 2020. https://standards.ieee.org/standard/2413-2019.html.

IEEE P2413, "Standard for a reference architecture for smart city (RASC)," IEEE Standard Association, 2019b. Accessed November 2020. https://standards.ieee.org/project/2413_1.html.

ISO, ISO 37120, Sustainable development of communities – indicators for city services and quality of life, 2014.

ITU-T, Focus group on smart sustainable cities. Smart sustainable cities: an analysis of definitions. Focus Group Technical Report, Geneva, Switzerland, Tech. Rep. FG-SSC–10/2014, 2014.

ITU-T, Recommendation Y.2060: overview of the Internet of Things, 2012. http://handle.itu.int/11.1002/1000/11559.

Khanna, A., Kaur, S., "Internet of Things (IoT), applications and challenges: a comprehensive review," *Wireless Personal Communications*, vol. 114, 2020, pp. 1687–1762. doi:10.1007/s11277-020-07446-4.

Larrucea, X., Combelles, A., Favaro, J., and Taneja, K., "Software Engineering for the Internet of Things," In IEEE Software, vol. 34 (1), 2017, pp. 24–28.

Lynn, T., Endo, P. T., Ribeiro, A. M. N. C., Barbosa, G. B. N., Rosati, P., "The Internet of Things: definitions, key concepts, and reference architectures." In: Lynn T., Mooney J., Lee B., Endo P. (eds) *The Cloud-to-Thing Continuum. Palgrave Studies in Digital Business & Enabling Technologies*. Palgrave Macmillan, Cham, 2020. doi:10.1007/978-3-030-41110-7_1.

Maciel, R. S. P., David, J. M. N., Claro, D. B., Braga, R., *Full Interoperability: Challenges and Opportunities for Future Information Systems. Grand Challenges in Information Systems for the Next 10 Years (2016-2026)*. SBC, Porto Alegre, 2017, pp. 107–118.

Manzano, W., Graciano Neto, V. V., Nakagawa, E. Y., "Dynamic-SoS: an approach for the simulation of systems-of-systems dynamic architectures," *The Computer Journal*, vol. 63, no. 5, 2020, pp. 709–731.

Maranhão, G. M., Bulcão-Neto, R., "A semantic filtering mechanism geared towards context dissemination in ubiquitous environments," *Journal of Universal Computer Science*, vol. 22, 2016, pp. 1123–1147.

Mauldin, T. R., Canby, M. E., Metsis, V., Ngu, A. H., Rivera, C. C., "Smartfall: a smartwatch-based fall detection system using deep learning," *Sensors*, vol. 18, 2018, p. 3363.

Mohanty, S. P., Choppali, U., Kougianos, E., "Everything you wanted to know about smart cities: the Internet of Things is the backbone," *IEEE Consumer Electronics Magazine*, vol. 5, no. 3, 2016, pp. 60–70.

Morin, B., Harrand, N., Fleurey, F., "Model-based software engineering to tame the IoT jungle," *IEEE Software*, vol. 34, no. 1, 2017, pp. 30–36. doi:10.1109/MS.2017.11.

Motta, R., de Oliveira, K., Travassos, G., "On challenges in engineering IoT software systems," *Journal of Software Engineering Research and Development*, vol. 7, Nos. 5:1–5, 2019, p. 20. doi:10.5753/jserd.2019.15.

Mulley, G., "Falls in older people," *Journal of the Royal Society of Medicine*, vol. 94, no. 4, 2001, p. 202.

Perera, C., Barhamgi, M., Bandara, A. K., Ajmal, M., Price, B., Nuseibeh, B., "Designing privacy-aware Internet of Things applications," *Information Sciences*, vol. 512, 2020, pp. 238–257, ISSN 0020-0255. doi:10.1016/j.ins.2019.09.061.

Perera, C., Barhamgi, M., De, S., Baarslag, T., Vecchio, M., Choo, K. R., "Designing the sensing as a service ecosystem for the Internet of Things," *IEEE Internet of Things Magazine*, vol. 1, no. 2, 2018, pp. 18–23. doi:10.1109/IOTM.2019.1800023.

Perera, C., Zaslavsky, A., Christen, P., Georgakopoulos, D., "Context aware computing for the Internet of Things: a survey," *IEEE Communications Surveys & Tutorials*, vol. 16, no. 1, 2014, pp. 414–454, First Quarter 2014. doi:10.1109/SURV.2013.042313.00197.

Santana, E. F. Z., Chaves, A. P., Gerosa, M. A., Kon, F., Milojicic, D. S., "Software platforms for smart cities: concepts, requirements, challenges, and a unified reference architecture," *ACM Computing Surveys*, vol. 50, no. 6, Article 78, 2018, 37 pp. doi:10.1145/3124391.

Santos, G. L., Monteiro, K. H. C., Endo, P. T., "Living at the edge? Optimizing availability in IoT." In: Lynn T., Mooney J., Lee B., Endo P. (eds) *The Cloud-to-Thing Continuum*. Palgrave Studies in Digital Business & Enabling Technologies. Palgrave Macmillan, Cham, 2020. doi:10.1007/978-3-030-41110-7_5.

Sheng, X., Tang, J., Xiao, X., Xue, G., "Sensing as a service: challenges, solutions and future directions," *IEEE Sensors Journal*, vol. 13, no. 10, 2013, pp. 3733–3741. doi:10.1109/JSEN.2013.2262677.

Shrivastava, R., Pandey, M., "Real time fall detection in fog computing scenario," *Cluster Computing*, vol. 23, 2020, pp. 2861–2870. doi:10.1007/s10586-020-03051-z.

Spinellis, D., "Software-engineering the Internet of Things," *IEEE Software*, vol. 34, no. 1, 2017, pp. 4–6. doi:10.1109/MS.2017.15.

Stevens, J. A., Mahoney, J. E., Ehrenreich, H., "Circumstances and outcomes of falls among high risk community-dwelling older adults," *Injury Epidemiology*, vol. 1, no. 1, 2014, p. 5. doi:10.1186/2197-1714-1-5.

Tanna, R. D., Kumar, K. S. V., Karthika, S., "Analytics as a service for beginners," *Proceedings of the 2017 International Conference on Computational Intelligence in Data Science (ICCIDS)*, Chennai, pp. 1–6, 2017, doi:10.1109/ICCIDS.2017.8272659.

Taylor, C., Siebold, A., Nowzari, C., "On the effects of minimally invasive collision avoidance on an emergent behavior." In: Dorigo M. et al. (eds) *Swarm Intelligence*. ANTS 2020. Lecture Notes in Computer Science, vol. 12421. Springer, Cham, 2020. doi:10.1007/978-3-030-60376-2_27.

United Nations, "68% of the world population projected to live in urban areas by 2050, says UN," 2018. Available at: http://tiny.cc/ow6ysz.

Weyrich, M., Ebert, C., "Reference architectures for the Internet of Things," *IEEE Software*, vol. 33, no. 1, 2016, pp. 112–116. doi:10.1109/MS.2016.20.

Zambonelli, F., "Key abstractions for IoT-oriented software engineering," *IEEE Software*, vol. 34, no. 1, 2017, pp. 38–45. doi:10.1109/MS.2017.3.

4 Smart Cities – *Energy*[1]

4.1 INTRODUCTION

At the energy level we are experiencing a trend that will change our traditional model, moving from centralized to decentralized power generation to produce wind and solar energy and reduce carbon emissions. This decentralized model is important given the limitations of storing wind and solar energy and therefore the need to produce them close to points of consumption. With this advancement, the importance of electrical energy is growing much faster than other forms of energy with more urbanization and, access to electricity electrification of many industries including transport.

As we get more digital, this increase in demand for energy in form of electricity will continue to grow, creating the greatest challenge in the history of humanity which is the sustainability of our future together with the exponential growth of technological development.

To make this happen, we need to make sure that the infrastructures are safe, intelligent and sustainable to decarbonize our model and improve the lives of new generations. This calls for decarbonization, decentralization, and digitalization of power generation.

This chapter will discuss how the Internet of Things digital technologies are changing practically all aspects of our lives and is transforming how we produce and consume energy with what is being called the 4th industrial revolution (Industry 4.0) with Communications between people, services, and things.

4.2 IoT FOR SMART ENERGY

The traditional electricity grid is characterized by large generation points of power plants that produce energy into the grid and transport it over kilometers to consumption points. This model is not sustainable when dealing with smart cities with distributed generation of renewable energy sources, thus requiring shifting the infrastructure investments from high voltage transmission to medium voltage distribution system that requires to balance the consumption with the smaller amount of electricity coming from distributed generation.

The success of the decarbonization of the energy model depends to a large extent on the capacity to integrate renewable sources in a massive way in each of the facilities or points of consumption, be it a home, a building or an industry. Every consumer of electricity should integrate renewable generation to reduce their energy costs and dependence on the traditional centralized grid, becoming a "prosumer" (producer and consumer). This means that power generation must gradually change from a centralized model in large power plants to a renewable generation model distributed throughout the entire electricity grid.

---

[1] This chapter was contributed by Mohamad Murywed.

DOI: 10.1201/9781003027799-4

Those intermittent renewable sources of energy (whether the wind is blowing or the sun is shining) must be coordinated together to guarantee the net balance of energy consumed and energy generated in the overall system. In other words, the electricity grid must be managed correctly to transfer excess energy from points where there is surplus renewable generation, to points where there is consumer demand and a limited or non-existent level of renewable generation due to unfavorable environmental conditions.

The solution to this challenge is the ability of any electrical installation to generate and transmit data in real time to the Electrical System Operator, to facilitate automation and its energy balance in a reliable way. Any electrical installation must be intelligent and capable of responding quickly to the needs of the grid, to weather conditions and to the control requirements of renewable generation. To do so, we need to introduce Internet of Things connected devices to digitalize any type of electrical installation and make them ready for the new network management model. This brings a new structural approach to generation and supply.

In addition, new technology like blockchain can be developed and integrated with IoT devices, for a seamless integration of information data bits to track the movement of energy electrons. This will bring transparency to the origin of energy sources (for example when utilities are marketing to consumers by stating that electricity is coming from 100% renewable source).

Another example where IoT is key in the sustainability of electrification is the need to store electrical energy. The facilities must be able to store the surplus energy that renewable sources have been able to generate at certain times of the day when the facility's consumption is lower. It does not make sense to waste these surpluses and use other polluting fuels to cover demand peaks at times when there are no favorable weather conditions for renewable generation. Therefore, we need IoT connected devices to better use the energy storage system (ESS) and uninterruptible power supply (UPS) system to be part of this new ecosystem to optimize the use and minimize the waste of energy as well as minimizing energy use at peak load times. This interoperability of IoT devices is one of the main pillars of IEEE P2030 Smart Grid working group.

One of those essential devices is Smart Meter or Advanced Metering Infrastructure which allows for Machine to Machine (M2M) interaction over the internet. This allows for real time reading of electricity demand and consumption which enables the decentralized energy system to respond to energy needs as well as promotion of energy efficiency giving the users visibility to energy use patterns. Prior to IoT connectivity, the manual process of periodically reading power consumption from analogue meters would only serve billing purposes and does not enable such data driven insights and actions.

This however comes with its cybersecurity risk just like any IoT application connected to the internet. Those Smart Meters must be secured (Hansen (2017)) to prevent intrusion of privacy or unauthorized access to data, as hacking into the readings can give an indication for example on what type of valuable electronics are at the house and make you vulnerable for malicious attacks when no one is home to plan for thieves breaking in. Cybersecurity must be looked at from a system point of view to prevent cyber attacks from the weakest link.

Another important point is data access and ownership while allowing seamless data integration into a variety of ecosystems. This is important to ensure that more data becomes available for use in the economy and society while keeping companies and individuals who generate the data in control. Some work needs to be done on what data the energy ecosystem requires and how will be facilitated.

On application side, this Smart Energy model is a key enabler for the transformation of personal mobility and transportation systems. This required a vehicle recharging infrastructure and new electrical installations that allow electric power to reach all recharging points. This requires technology to protect these new infrastructures, and to connect them to all types of IoT systems to provide the best service to the user. For example, cargo payment systems, or digital maintenance services to avoid service stoppages.

This charging infrastructure when installed at a building, can also serve in having the electric vehicle act as an additional battery when there is sudden additional demand for electricity enabled by the EV to grid IoT devices. This adds stronger need for Smart Home/Building, by automating them for an optimal cost-effective use of energy based on availability and storage of energy. IoT technology is required to protect the power grid from EV overload.

In parallel, with the rise of load from EV charging, IoT becomes essential to manage and reduce the load from industrial motor-based applications. According to International Energy Agency (IEA), the electric motors and systems they drive are the largest energy users and account for more than 40% of global electricity consumption (IEA, 2011). With IoT enabled Wireless Smart Sensors that can be easily installed on Motors and act just like a Smart Watch giving an indication on speed, energy consumption, and health status (temperature, noise, vibration), which can be used to optimize the maintenance and operations in a more proactive predictive manner as well as a more efficient and continuously monitored way to save energy and load on the system. To take this a step further some older low efficiency motors can be retrofitted with addition of power electronics that change the mechanism from fixed to variable speed leading to substantial energy savings. To give you an idea on how older motors operated, think of driving a car where you have to step on gas pedal to the maximum and at the same time keep applying brakes to get to the desired speed, wasting most of the fuel. With introduction of power electronics, you can precisely control the speed of acceleration and deceleration with gas pedal versus mechanical brakes. Similar type of system can be implemented to older motors as well as automating speed control with IoT sensors.

Beyond the residential chargers for electric cars, we need IoT to transform the collective transport model of public and private entities, with infrastructure for fleets of electric buses that require intelligent and fast charging systems. This type of infrastructure requires digital technology to connect devices to remote control centers and digital services to optimize their maintenance.

The additional amount of data generated by such IoT applications highlights the importance of reliable data centers. Think about all the exponential growth of devices connected to the Internet and all the servers that process and maintain data 24 hours a day, 365 days a year, for all types of applications. The power supply of these servers is critical and requires a totally reliable, redundant and intelligent infrastructure that allows ensuring continuous uninterrupted service with IoT connected smart panels and uninterruptible power systems (UPS). Those devices enable monitoring of all parameters of the installation in real time, supervise its condition in a proactive predictive way and automate it to guarantee reliability. Automation will further develop into autonomous self-managed operations with the advancement of Artificial Intelligence using all the data generated by those IoT connected devices.

One of the challenges we have with IoT, is going beyond the digitalization of a product, a system or a process. What is relevant is what to do with the data they generate. Digitization with IoT should serve to generate useful data and process it to solve a specific user problem in an easier, faster and more accessible way.

There is a lot of work yet to be done on creating digital platforms and ecosystems that give more visibility of processes, receive information in a more orderly and intuitive way, discover new possibilities to improve and seamlessly integrate artificial Intelligence and Machine Learning algorithms to provide incremental value. For this to happen, we need open yes secure systems that allow for interoperability of all the customizable applications.

IoT information technology (IT) is critical for the optimization of Industry operational technology (OT) for a faster and more agile production system that can help with the changing demands and further customization needs of consumers. 2020 pandemic has shown us the growing need for agility and flexibility that can be best achieved with Smart Energy. At the operational level with more mobile workforce there is a need to increase the safety of the installation and reduce operating costs related to the use of energy and the management of the maintenance of assets. With increasing remote work culture, there is an exponential increase in need for IoT for automated load control and early detection of possible failures to prevent outages or stopped production.

Access to data from IoT devices would increase the transparency of all your processes and their possible inefficiencies. This would enable the management of data of the installation in a flexible way and for a better and more automated decision making and analysis. After all, data generated by IoT is the fuel of digital transformation.

4.3 EXAMPLES OF OVERCOMING SMART ENERGY CHALLENGES

4.3.1 GORONA DEL VIENTO WIND-HYDRO AT EL HIERRO CANARY ISLAND IN SPAIN

One of the challenges of Smart Energy is the current cost of storage using lithium ion batteries. A Smart project to overcome this limitation has been done by Endesa Gorona del Viento (https://www.endesa.com/en/projects/all-projects/energy-transition/renewable-energies/el-hierro-renewable-sustainability) at El Hierro Canary Island in Spain. They built a wind farm with enough capacity to meet 100% of electricity demand of the island. The surplus wind energy which is not consumed by the Island's population is used to pump water from a lower reservoir to a higher reservoir. This way water acts as the energy storage medium that is later used to generate power using hydro power turbines when wind production is low. The combination of wind and hydro power successfully transforms an intermittent energy source like wind into a continuous reliable source with the help of digital smart meters installed at consumption points that can indicate the real time demand. This has helped the island move from expensive and polluting fuel-based power generation to 100% clean renewable energy.

4.3.2 VIRTUAL POWERPLANT AND MICROGRID AT ARUBA CARIBBEAN ISLAND

Virtual Powerplant advanced control system is used at Aruba island Endesa Gorona del Viento (https://www.endesa.com/en/projects/all-projects/energy-transition/renewable-energies/el-hierro-renewable-sustainability) as it does not have natural ways to store the energy as per El Hierro island. Giving examples on islands as it is easier to visualize but this applies to all type of locations. With Smart Energy, power is carried across many interconnected islands "microgrids". With advanced automation and control technologies the electricity generation system can function as a virtual power plant integrating variable resources like wind and solar with the help of weather forecasting and real time monitoring to meet the demand while keeping the grid stable and using less power generation from traditional fossil fuels. With this predative method of energy management, additional loads coming from electric mobility can be accommodated and incentivize people to charge at home with lower electricity prices when there is abundance of renewable energy sources. This will lead to wider adoption of electric vehicles and further lower the pollution and fossil fuel dependence.

4.4 CONCLUSION

As we have seen that Smart Energy is essential and no longer optional for us to co-exist and flourish with nature without consuming the planet. We need to produce and store energy in a sustainable and cost-effective way with digitalization to associate the bits with electrons in the decentralized system for decarbonization of the world to make zero-emissions a reality. Some aspects like storage, cybersecurity, and data ownership need big attention in the development.

FURTHER READING

Hansen, A., Staggs, J., Shenoi, S. "Security analysis of an advanced metering infrastructure," *International Journal of Critical Infrastructure Protection*, 2017. https://www.sciencedirect.com/science/article/abs/pii/S1874548217300495?via%3Dihub

IEA on Energy Efficiency Policy Opportunities for Electric Motor-Driven Systems. https://www.iea.org/reports/energy-efficiency-policy-opportunities-for-electric-motor-driven-systems, 2011.

IEEE P2030.9. Guide for smart grid interoperability of energy technology and information technology operation with the electric power system (EPS), end-use applications, and loads. https://standards.ieee.org/project/2030.html, 2019.

5 Smart Cities – *Security*[1]

5.1 INTRODUCTION

Smart cities use networked technology to gather data to optimize and manage city resources, businesses, and government entities. The optimization of a smart city comes from leveraging the data aggregated and analyzed that is collected from a city's assets. With the appropriate analysis of data collected from a city's assets, city components such as traffic patterns, energy distribution, and air quality can be improved more efficiently. These improved elements of a smart city can also increase the quality of life and convenience for its citizens, by integrating IT technology into many aspects of a citizen's day-to-day tasks such as simplifying access to many of a city's services. According to the American Society of Mechanical Engineer (ASME), subjects such as mobility, healthcare, water, energy, engagement and community, economic development and housing; waste, and security can all be integrated into the core of a smart city to create an infrastructure that makes a city smart (Kosowatz, 2020). The top three US smart cities are New York, NY; Boston, MA; and San Francisco, CA (Datalux.com). Smart cities have also been deployed in many places such as Europe, Singapore, and South Korea.

Smart city implementations come with challenges such as insufficient funds, inconsistent network connectivity, lack of experienced professionals, and security risks (Joshi, 2020). The focus of this chapter will be on cybersecurity, and preparing countermeasures as the cyberattacks in a smart city can threaten the lives of its citizens. Smart city security threats will be examined from a multilayer perspective, targeting representative elements that make up a smart city. This chapter will also describe attack scenarios and threat countermeasures. In addition, a discussion on educating the professionals needed to design and build smart cities will be provided.

5.2 SMART CITY ARCHITECTURE

One of the goals of a smart city is to improve the quality of life of people by building an IT infrastructure that improves the many pieces of everyday life. Figure 5.1 shows the multilayer architecture of a smart city. Figure 5.1 also shows the multiple attack points of each layer. Security and privacy are essential in a smart city environment as many end-point devices are connected to the Internet, communicate with each other, and accumulate data stored in a cloud. Cyberattacks on a smart city can debilitate critical system such as water and electricity. In addition, these attacks can be very damaging on an individual level because smart cities are enabled by cyber-physical systems (CPSs), which involve connecting devices and systems such as Internet of Things (IoT) devices (Aldairi & Tawalbeh, 2017). Emergency alert systems, street surveillance, and smart traffic lights are very vulnerable (Lawrence, 2021).

---

[1] Excerpts of this chapter are from Chung, Park and DeFranco (2021).

DOI: 10.1201/9781003027799-5

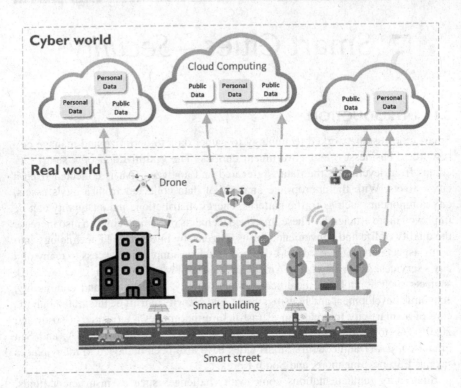

FIGURE 5.1 Smart city architecture.

To appreciate the severity, recall the 2018 situation in Hawaii where the emergency management agency mistakenly sent out this message,

> BALLISTIC MISSILE THREAT INBOUND TO HAWAII. SEEK IMMEDIATE SHELTER. THIS IS NOT A DRILL.

This false alarm was apparently not caused by hackers (according to the news), nonetheless critical phone systems have had many attacks – where the attackers are looking for ransom payments in addition to determining where the weak points are in a critical network (Starks, 2018).

Researchers have presented security and privacy challenges for intelligent healthcare, transportation, smart building, and smart energy (Braun et al., 2018). However, there have been no studies on systematically addressing security with a multilayer structure to include device/sensor level, network level, application level, and cloud level from viewpoint of security and privacy. In the remainder of this chapter, we will look at the attack points and countermeasures that exist on multiple layers of a smart city.

5.3 ATTACK POINT EXAMPLES IN A SMART CITY

In this section, we will discuss each of the attack point levels (i.e., device/sensor, application, network, cloud/data) with four smart city elements: smart transportation, smart healthcare, smart energy, and smart buildings.

TABLE 5.1

The Device/Sensor Attack Points in Smart City

Smart City Element	Device/Sensor Level
Smart transportation	Smart car
	Street sign
	Smart traffic lights
	Streetlights
	Smart parking sensor
Smart healthcare	Smart watch
	Fitness tracker
	Smart pacemaker
	Closed-loop insulin delivery (artificial pancreas)
Smart energy	Air-conditioning
	Heating sensor
	Water leakage sensor
	Light sensor
	Temperature and humidity sensor
Smart building	Smart thermostat
	Smart camera
	Smart speaker
	Smart door lock
	Smart baby monitor

5.3.1 DEVICE/SENSOR LEVEL

Hardware chips of hub devices such as those that control digital devices or sensors (Table 5.1) can be exploited by physical attacks. Sensors used by smart cities are often built into traffic lights and streetlamps to manage parking and traffic congestion; however, most wireless sensors are a vulnerability risk because they do not have built-in security. In addition, there are cases where the smart devices used in traffic lights, healthcare devices, energy sensors, and home environments are attacked at the hardware level (The 8 Most Common Types of Cyber Attacks Explained, 2019). For example, there were vulnerabilities discovered in Medtronic's infrastructure that enabled an attacker to control an implanted pacemaker remotely due to compromised software updates (Newman, 2018).

5.3.2 APPLICATION LEVEL

As shown in Table 5.2, each element of smart city uses mobile or web applications to control the devices/sensors and to manage the data generated. Applications are a frequent target of cyberattacks via cross-site scripting (XSS) attacks (e.g., injecting malicious script into a client) or man-in-the-middle (MITM) attacks (e.g., data transfer eavesdropping). Smart cities are also a target of these type of attacks where transmitted data can be intercepted or altered.

TABLE 5.2

The Application-Level Attack Points in Smart City

Smart City Element	Application Level
Smart transportation	Website, application for desktop (Windows, macOS),
Smart healthcare	application for mobile (iOS, Android)
Smart energy	
Smart building	

Websites are also at risk. Since they are at the application level and are indispensable to control the devices or systems that make up a smart city, they can be a first attack point for a malicious actor. For example, in 2017, an XSS attack occurred on eBay where malicious JavaScript was injected onto auction descriptions. This vulnerability was exploited and inserted redirection code to a list of expensive vehicles (Mutton, 2017).

5.3.3 NETWORK LEVEL

The devices, sensors, hubs, interfaces, and clouds that are part of the smart city system communicate with each other. For communication, there are various network protocols, as shown in Table 5.3, to enable short-range communication (Garcia et al., 2018). Possible cyberattacks that can occur at the network level include MITM attacks and packet tampering that are used to capture and/or alter data. For example, Samsung smart refrigerators and other smart home devices have been the target of MITM attacks where hackers infiltrate the home network and steal the device owner's e-mail credentials (Voxill Tech).

5.3.4 EDGE/CLOUD LEVEL

At the edge/cloud level, the devices or sensors, shown in Table 5.4, are examples that partially make up the smart city. Those sensors generate and send digital data in real time to the cloud, where data analytics is performed to help city planners make informed decisions for improving a city's function. However, large amounts of data stored remotely can be a risk. In a smart city, data generated from people,

TABLE 5.3

The Application-Level Attack Points in Smart City

Smart City Element	Network Level
Smart transportation	Internet (4G, 5G), Wi-Fi, Zigbee,
Smart healthcare	Zwave, Bluetooth, WSN, LoRa
Smart energy	
Smart building	

TABLE 5.4

The Application-Level Attack Points in Smart City

Smart City Element	Cloud/Data Level
Smart transportation	Public traffic, vehicle recording, signal control
Smart healthcare	Health data, GPS, medical records
Smart energy	Customer location, temperature, humidity, power quality
Smart building	Human behavior, GPS, user voice, user interest, movement, photo/video

organizations, and government agencies are all stored remotely, and if the data is stolen, privacy infringement occurs and be used by the attackers for financial gain (Elmaghraby & Losavio, 2014). Given cloud data storage plays a key role in smart cities, data breaches at the cloud level can cause serious damage (Pedigo).

5.4 THREAT SCENARIOS

Clearly smart city cyberattacks are a reality. In this section, attack scenarios are discussed. The threat scenarios are presented in Table 5.5 in accordance with the attack techniques that can occur at each of the attack points (Kitchin & Dodge, 2019). In Table 5.5, the threat scenarios are mapped to the smart city attack points to the MITRE ATT&CK Matrix, which is a global database of adversary tactics (MITRE, ATT&CK Framework). This table highlights the prevalent smart city attack techniques and tactics.

Table 5.6 maps the same smart city threat scenarios (i.e., traffic, medical, energy, building, and privacy) to ENISA's Cybersecurity Smart City Architecture security parameters (Lévy-Bencheton, 2015). These security parameters address/model the *confidentiality* (i.e., protecting information from being accessed by unauthorized actors), *integrity* (i.e., ensuring data is not altered), and *availability* (i.e., data is accessible by authorized users) (a.k.a. CIA Model or CIA Triad) to the corresponding threat scenario. In the next few sections, each threat scenario will be discussed in further detail.

5.4.1 TRAFFIC CHAOS

City transportation hacks are the most prominent attack in smart cities as autonomous driving and unmanned shuttle services, smart parking services, and other vehicle communications are normal. Hackers attack autonomous driving shuttles by taking control of the vehicle in an attempt to cause unexpected traffic chaos. Attackers gain control of the vehicles by creating and modifying communication messages. For example, using security vulnerabilities, US security company IOActive successfully hacked a major urban traffic control system (Cerrudo, 2015). This demonstrated the traffic chaos a hacker can cause.

Examples at the device or sensor level are supply chain compromise, pre-OS boot, and firmware corruptions. In applications that control the transportation system, attacks using credential access can occur as well as attacks focusing on data

TABLE 5.5
Possible Threat Scenarios in Smart Cities

Attack points	ATT&CK tactic	ATT&CK technique	Possible Threat scenario				
			Traffic	Medical	Energy	Building	Privacy
• Device/sensor	• Initial Access	• Supply chain compromise	X	X	X	X	
	• Persistence	• Pre-OS Boot	X	X	X	X	
	• Impact	• Firmware corruption	X	X	X	X	
• App.	• Credential Access	• Brute force	X	X	X		X
		• Credentials from password stores	X	X	X		X
	• Impact	• Data destruction	X	X	X		
		• Data encrypted for impact	X	X	X		
		• Endpoint denial of service (DoS)	X	X	X		
		• MITM	X	X	X		
• Network	• Credential Access	• Network sniffing	X	X	X	X	
	• Impact	• Network DoS	X	X	X	X	
		• Service stop	X	X	X		
• Edge/cloud	• Exfiltration	• Exfiltration over other network medium					X
		• Scheduled transfer					X

TABLE 5.6

Proposed Scenarios and Security Parameters

Threat scenario	Security Parameters
• Traffic chaos	• Availability
• Medical ransomware	• Integrity
• Energy system hacking	
• Building attack	• Authenticity
	• Integrity
• Privacy	• Confidentiality

destruction. In addition, the devices that make up the smart transportation system are networked to each other, which can lead to MITM, network denial-of-service (DoS), and service stop attacks.

5.4.2 MEDICAL RANSOMWARE

Management of individual medical services, remote medical examinations, and the use of smart medical devices are becoming more prevalent in smart cities. Thus, smart medical devices, applications, medical information, Big Data systems, and medical data management are major cybercrime targets. If a medical device that is directly connected to an individual, such as a respirator or closed-loop insulin pump system is hacked, the patient's life is at risk. Personal patient information is also at risk. In 2018, a medical group's database was infected with malware, and about 16,000 prescriptions were leaked, including the local Prime Minister (Davie, 2018). By exploiting these attacks, ransomware is often used to exploit money from patients or their caregivers.

Attacks such as supply chain compromise, pre-OS boot, and firmware corruption are also likely to be carried out at the medical equipment, device, or sensor level. In order to gain access to the hospital's application that manages medical equipment or to the patient's application, hackers can attack to steal credentials and perform data destruction and data encryption. Devices, medical institutions, and patients that make up smart medical equipment are networked to each other, which can lead to frequent network-level attacks such as MITM attacks, network DoS attacks, and service stop attacks. If this happens, it could pose a major threat to patients and medical institutions.

5.4.3 ENERGY SYSTEM HACKING

Cyberattacks aimed at energy collection sensors and management systems are also a tremendous risk to smart cities. In the Ukraine, power grids were hacked in 2015, causing 30 substations to cease operation and the electricity supply to suspend its operation. In April 2016, a power plant facility in Michigan, USA, was infected with ransomware (Condliffe, 2016). This attack was aimed at a programmable automatic control unit (PLC), which is a significant part of the operation of a factory as it can have fatal consequences. If the PLC of chemical factories is transferred to hackers, the neighborhood or the country is at risk – as this control produces a substance that is harmful to the human body.

At the hardware level, in the device itself that constitutes the energy system, attacks such as supply chain compromise, pre-OS boot, and firmware corruption are a risk. Attackers targeting software that manages energy systems can attempt to exploit credentials and initiate attacks that involve data destruction and data encryption for impact. The possible attacks are MITM, network DoS, and service stop attacks that can occur over the network within a networked energy system.

5.4.4 BUILDING ATTACK

When a house or organization is infected with malware, the malware can be transmitted through smart device communication. If a hacker attacks an area of a building successfully, it can be said to have the potential to infect the entire organization. Residents can be filmed/recorded at home by hacked smart TVs and infected AI speakers can eavesdrop on personal conversations. In addition, smart video conferencing devices can record the content of meetings between employees in the company. These are all examples of privacy violations.

At the hardware level, which is the device itself that makes up a smart home or smart office, attacks such as supply chain compromise, pre-OS boot, and firmware corruption are likely to occur. In this environment, as in other scenarios, attacks on the network, for example, MITM, can occur because it is a networked environment. In addition, since smart homes or offices are relatively narrow spaces, data can be stolen from the network or prevent devices from operating properly through signal jamming attacks.

5.4.5 PRIVACY BREACH

Hacking a data system in a smart city that uses Big Data to provide citizens with various convenient services is on a much larger scale of risk due to private information leakage. Cyberattacks on publicly sensitive data such as city administration, healthcare, and finance can be devastating to city operations. A perfect example is in 2017, a database of hospital network 'Atrium Health' was hacked in the United States, and 2.65 million personal records were leaked (Miliard, 2018). When a user's credentials are hijacked, a security break-in can occur as the key required to access the data is acquired. There is also a possibility that attack techniques related to data capture at the edge/cloud level may be applied.

5.5 SECURING SMART CITIES

There are many ways to secure smart cities. Possible strategies include the following and will be discussed further in the upcoming sections:

- *Cyber Risk Management*: Urban planners and smart city governments can proactively seek ways to make their cities' infrastructures safe from potential cyberthreats.

- *Cyber Patrol Bots*: Capture data on the overall state of the system, including end-point devices and connectivity traffic.
- *Security and Privacy Labels*: Security and privacy labels for huge systems as well as single IoT products that make up a smart city should be provided with detailed descriptions of cybersecurity risks.
- *Education*: Make available education programs to cultivate talent and skilled cybersecurity experts (Kitchin & Dodge, 2019).

5.5.1 CYBERSECURITY RISK MANAGEMENT

Urban planners and smart city governments need to proactively seek ways to make their cities' infrastructures safe from potential cyberthreats. This can be done by managing the security needs of the city at multiple levels as explained earlier. It is also necessary to analyze the security threats of the smart city components and establish protection measures for each layer according to the multilevel measures described above.

A smart city should consider a plan of action using the National Institute of Standards and Technology (NIST) Cybersecurity Framework to improve the management of cybersecurity risk (NIST, 2018). The framework provides detailed guidance to help develop a cybersecurity profile in order to prioritize and align cybersecurity activities with city risk tolerance and requirements.

Furthermore, smart city designs need to meet legal requirements such as outlined in the 'Personal Information Protection Act' (governs how privacy organizations are allowed to collect, use, and disclose personal information in the commercial business space – Canada) and the 'Geolocation Privacy and Surveillance Act' (establishes a framework/guidelines for agencies, entities, and cities to know how and when geolocation information can be utilized and accessed – USA). All services must be designed and developed by considering the information protection requirements.

5.5.2 CYBER PATROL BOT

Just as police patrol monitor dangerous/vulnerable areas, it is necessary to monitor smart technology to prevent and detect attacks. A bot is needed to automatically detect and patrol abnormalities through data-generating devices and sensors in the smart city operation. Cyber patrol captures data on the overall state of the system, including endpoint devices and connectivity traffic. This data is then analyzed to detect possible security vulnerabilities or potential system threats. Once detected, a broad range of actions are formulated in the context of an overall system security policy that should be executed, such as quarantining devices based on behavior anomalies.

The solution has several limitations. The cyber patrol bot will be effective in situations where various systems are interoperable. Thus, in order to implement this solution in the real world, it will only be possible for uniform systems. Moreover, it can be difficult to be interoperable with state-of-the-art systems even for legacy systems such as the traffic management systems that can be somewhat antiquated.

5.5.3 Security and Privacy Label

Most consumer products are labeled with a description of the product for safety reasons. On groceries, the ingredients and processing plant information/location are provided to the consumer. Food nutrition labels were developed to decrease obesity and mitigate medical conditions such as severe food allergies and diabetes who need food carbohydrate and sugar information to dose medication.

Similar to this concept, a label for IoT devices, sensors, and systems that makes up a smart city should be provided with detailed descriptions of the security and privacy risks. The concept of labeling a single product was proposed by Emami-Naeini et al. (2020). In a smart city, security and privacy labels for huge systems as well as single IoT products should be delivered to the administrator in the form of step-by-step instructions or notification windows.

Specifically, to summarize the information to be included on a label:

- Security mechanisms
  - Security update dates (e.g., update available January 1, 2021)
  - Access control passwords
- Data practices:
  - Sensor data collection (e.g., text, video, audio)
  - Data stored on the device or edge/cloud that is shared.

5.5.4 Nurturing Talented Cybersecurity Personnel

For the security of smart cities, an education program that can cultivate the best talent should be established. When developing a smart city project, local governments will face the same workforce issues familiar to everyone in the private sector: shortage of talent/workforce with greater cybersecurity skills. This only increases the security concerns inherent in these projects. To make cities safer, countries and cities need to cooperate with schools and institutions to cultivate new workforces and systematically create educational programs to reeducate existing workforces. A complete CPS/IoT curriculum requires a skill set from existing engineering and computer science programs. To begin with, it is recommended to start with developing IoT/CPS elective courses in these major academic programs or at the very least, add CPS/IoT concepts to existing courses in those programs (DeFranco, Kassab & Voas, 2018). More on this topic is included in the IoT education chapter of this book.

5.6 CONCLUSION

The smart cities are being built and expanded. Therefore, we must be vigilant to keep the citizens safe from nefarious actors looking for vulnerabilities in these systems. There are possible attacks for each element, device, and sensor that make up the smart city. In this chapter, the vulnerabilities and possible security solutions for each network layer were presented by dividing them into hardware, application, network, and cloud levels. Based on these attack points, proposed attack scenarios and countermeasures to counter those threats were discussed. These topics are important and are what is necessary to create a SAFE smart city.

FURTHER READING

Aldairi, A., Tawalbeh, L., "Cyber security attacks on smart cities and associated mobile technologies," *Procedia Computer Science*, vol. 109, 2017, pp. 1086–1091. doi:10.1016/j.procs.2017.05.391.

Braun, T., Fung, B. C. M., Iqbal, F., Shah, B., "Security and privacy challenges in smart cities," *Sustainable Cities and Society*, vol. 39, 2018, pp. 499–507. doi:10.1016/j.scs.2018.02.039.

Cerrudo, C., "An emerging US (and World) threat: cities wide open to cyber attacks," White paper, 2015. https://ioactive.com/pdfs/IOActive_HackingCitiesPaper_CesarCerrudo.pdf.

Chung, H., Park, J., DeFranco, J., "Multi-layered diagnostics for smart cities," *Computer*, 2021.

Condliffe, J., "Ukraine's power grid gets hacked again, a worrying sign for infrastructure attacks," *MIT Technology Review*, 2016. https://www.technologyreview.com/2016/12/22/5969/ukraines-power-grid-gets-hacked-again-a-worrying-sign-for-infrastructure-attacks/.

Datalux.com, "Top 10 smart cities in the US,", 2018. https://datalux.com/top-ten-smart-cities-us/#:~:text=Top%20Ten%20Smart%20Cities%20in%20the%20U.S.%201,Los%20Angeles%2C%20CA.%20...%2010%20Atlanta%2C%20GA.%20.

Davie, J., "The 10 biggest U.S. healthcare data breaches of 2018," *Health IT Security*, 2018. https://healthitsecurity.com/news/the-10-biggest-u.s.-healthcare-data-breaches-of-2018.

DeFranco, J., Kassab, M., Voas, J., "How do you create an Internet of Things workforce?," *IT Professional*, vol. 20, no. 4, 2018, pp. 8–12.

Elmaghraby, A. S., Losavio, M. M., "Cyber security challenges in smart cities: safety, security and privacy," *Journal of Advanced Research*, vol. 5, no. 4, 2014. doi:10.1016/j.jare.2014.02.006.

Emami-Naeini, P., Agarwal, Y., Cranor, L. F., Hibshi, H., "Ask the experts: what should be on an IoT privacy and security label?," *IEEE Symposium on Security and Privacy (SP)*, San Francisco, CA, 2020.

Garcia, L., Jimenez, J. M., Taha, M., Lloret, J., "Wireless technologies for IoT in smart cities," *Network Protocols and Algorithms*, vol. 10, 2018. doi:10.5296/npa.v10i1.12798.

Joshi, N., "Four challenges faced by smart cities," BBN Times, November 24, 2020. https://www.bbntimes.com/technology/four-challenges-faced-by-smart-cities.

Kitchin, Rob, Dodge, M., "The (in)security of smart cities: vulnerabilities, risks, mitigation, and prevention," *Journal of Urban Technology*, vol. 26, no. 2, 2019, pp. 47–65.

Kosowatz, J., "Top 10 growing smart cities," American Society of Mechanical Engineers, February 3, 2020. https://www.asme.org/topics-resources/content/top-10-growing-smart-cities.

Lawrence, C., "Which Smart City Tech is most vulnerable to cyberattack?" IoT Techtrends, April 5, 2021. https://www.iottechtrends.com/what-smart-city-tech-is-vulnerable-to-cyberattack/.

Lévy-Bencheton, C., *Cyber Security for Smart Cities: an Architecture Model for Public Transport*. ENISA, Heraklion, 2015.

Miliard, M., "Atrium Health breach: data from 2.65 M patients potentially exposed," Healthcare IT News, 2018. https://www.healthcareitnews.com/news/atrium-health-breach-data-265m-patients-potentially-exposed.

MITRE, ATT&CK Framework, 2021. https://attack.mitre.org.

Mutton, P., "Hackers still exploiting eBay's stored XSS vulnerabilities in 2017," February, 2017. https://news.netcraft.com/archives/2017/02/17/hackers-still-exploiting-ebays-stored-xss-vulnerabilities-in-2017.html, retrieved 9/20/2020.

Newman, L., "A new pacemaker hack puts malware directly on the device," Wired, September 18, 2018. https://www.wired.com/story/pacemaker-hack-malware-black-hat/, retrieved 9/10/2020.

NIST, "Framework for improving critical infrastructure cybersecurity," April 16, 2018. https://nvlpubs.nist.gov/nistpubs/CSWP/NIST.CSWP.04162018.pdf.

Pedigo, C., "The biggest Cloud Breaches of 2019 and how to avoid them for 2020," December 13, 2019. https://www.lacework.com/top-cloud-breaches-2019/, retrieved 9/20/2020.

Starks, T., "Hawaii missile alert highlights hacking threat to emergency systems," Politico, January 16, 2018. https://www.politico.com/newsletters/morning-cybersecurity/2018/01/16/hawaii-missile-alert-highlights-hacking-threat-to-emergency-systems-074411.

The 8 Most Common Types of Cyber Attacks Explained. 2019. https://www.comtact.co.uk/blog/common-types-of-cyber-attacks-explained, retrieved 9/10/2020.

Voxill Tech, "Samsung smart fridge vulnerability can expose Gmail credentials", Medium, April 4, 2017. https://medium.com/@voxilltec/samsung-smart-fridge-vulnerability-can-expose-gmail-credentials-says-experts-1a992cb751de, retrieved 9/10/2020.

6 Smart Homes[1]

6.1 INTRODUCTION

The smart home (SH) market is exploding. In fact, the market is predicted to increase to \$174 billion by 2025 (Marr, 2020). The popularity stems from the many innovative SH devices available to make our lives more comfortable, safe, and convenient – such as devices that monitor a person's health (e.g., monitor sleep, track heart activity), increase safety (e.g., alert caregivers), and some that learn and adapt to the preferences and behavior of the homeowners. This is all possible, of course, by the advancement of engineering, increased processing speed, powerful networks, data analytics, artificial intelligence (AI), etc.

As engineers, we know that a SH is much more complex than adding a self-stocking beer refrigerator or an intelligent virtual assistant (IVA) such as Amazon Alexa, Google Home, and Apple's Siri to play your favorite music, open the blinds, order more paper towels, make dinner reservations, etc. The complexity comes from the required interoperability of all the devices and systems in the home and the need for the entire system to be energy efficient and ensure the privacy and safety of their inhabitants. That said, engineers and developers need to understand everything about SH systems to keep improving the security, privacy, scalability, performance, interoperability, efficiency, and usability of SH systems.

The major themes of SH design and development among researchers are as follows:

- *Security Design and Management*: This area includes security systems and management, device security, risk management, security architecture, application security, intrusion detection, encryption, authentication, and privacy.
- *Products*: This topic includes specific products for home comfort, such as emotion/social connectedness, gardening assistance, entertainment, applications to control household devices, home automation systems, smart house assistants, and health notification/monitoring.
- *Activity and Behavior Patterns*: This area is focused on identifying human activity and behavior patterns, models for planned behavior, and user location and discovery (ULD).
- *Power Efficiency*: Developers in this area focus on device/system power consumption/efficiency and energy optimization.
- *Systems Design, Simulation*: This area includes systems design, models, simulations, requirements, architectures, frameworks, cost models, performance-improving algorithms, and interoperability solutions.

---

[1] Contributions to this chapter were made by Nina DeFranco Tommarello and Hyunji Chung.

DOI: 10.1201/9781003027799-6

Next, each of these SH research themes is explained in further detail. This chapter concludes with a description of two do-it-yourself SH devices: a smart garden and a smart trashcan.

6.2 SECURITY DESIGN AND MANAGEMENT

Security is one of the most predominant SH research areas. When designing a SH, there are two main categories of home security to consider: cyber and physical. Cybersecurity can be more of a challenge to manage because for every device added to the SH, there is an added way for a hacker to break into the system. This is not to minimize the importance of the physical security of a SH – but now with Internet of Things (IoT), activity monitoring, facial recognition, and real-time notifications can be added to the traditional security system.

6.2.1 SMART HOME SECURITY OVERVIEW

Security should be a consideration in the initial design of all SH products. Some products are more vulnerable than others such as those that feature multiple users, use web-based services, and have multiple in-home sensors. The challenge is that these products need to manage security while also improving communication and reliability (Golvilkar et al., 2018).

Specific product examples focusing on security are those that enable real-time monitoring of the home for motion and disturbance detection, automation of lighting, temperature, and humidity of the home (Sharma et al., 2015). Researchers working on security design and management focus on the best practices to mitigate threats, vulnerabilities of the SH system. In other words, the approaches to integrate security in SH designs include standards and best practices to deal with security threats and the countermeasures to secure vulnerabilities in SH design (Batalla, Vasilakos & Gajewski, 2017). Typical threats include privacy and security attacks where assets are compromised, physical attacks on smart devices where device firmware and hardware settings are manipulated, and denial of service.

Direct consequences of a cyberattack on a SH include financial, vocational, invasion of privacy and inconvenience, and much more. To help researchers from diverse disciplines evaluate SH attacks, a classification of SH security threats to highlight vulnerabilities of SH configurations was identified (Heartfield et al., 2018):

- *Cyber Threat Impact*:
  - Confidentiality (e.g., keeping sensitive data private)
  - Integrity (e.g., accuracy, completeness, and consistency of SH data)
  - Availability (e.g., data is available at a required level of performance)
  - Nonrepudiation (e.g., assurance that the data sender is provided with proof of delivery AND the data recipient is provided with proof of the sender's identity)
- *Physical Threat Impact*:
  - Unauthorized actuation (i.e., an authorized user did not initiate or approve the action to cause something in the home to operate).

- Incorrect actuation (i.e., the operation was not as required by an authorized user).
- Prevented actuation (i.e., authorized user not able to initiate a desired actuation).
- Delayed actuation (i.e., actuation is initiated or completed later than desired by the user).
- Breach of physical privacy (i.e., covertly recording audio in a home).

Another category of home security focuses on the inside of the home autonomously detecting specific security-related situations (Dahmen et al., 2017):

- *Detecting Resident-Based Target States*: Systems using sensors learn inhabitant activity/inactivity patterns throughout the home to determine if intervention is needed. Another area of research is to detect individual falls with the aging population. There are many methods using audio and video sensors, wearable sensors, and vibration sensors in the floor.
- *Detecting Home-Based Target States*: Sensing the location of inhabitants, looking for anomalies (i.e., finding data not conforming to expected behavior), and sensing visitors or intruders.
- *Home-Based Anomaly Behavior Detection*: Techniques to determine unexpected behavior patterns. Location in the home using heat mass doors and window sensors, etc.
- *Activity Anomaly Detection*: A deeper level of activity recognition (i.e., health monitoring such as sleep patterns or a decline in cognitive health).

6.2.2 Smart Home Design and Management

Security and privacy go hand and hand. A security breach causes a vulnerability that could expose an unsuspecting user to a privacy breach. For example, an IVA has four possible attacks: wiretapping, direct communication to an IVA-enabled device, malicious voice commands, and unintentional voice recordings. Figure 6.1 represents the four possible attacks in the IVA environment.

The IVA ecosystem is complex and heterogeneous, so there are also various points where attacks can occur. In the first scenario, the IVA-enabled device is an attack that intercepts conversations between people and speakers by wirelessly applying the protocols that communicate with the cloud. In the second scenario, the speakers that are compromised by hackers are eavesdropping devices that record what users say. Furthermore, hackers may surprise users by playing songs or grotesque sounds on speakers. In the third scenario, the hacker impersonates the user as a starting point for attacking the smart home to carry out criminal acts such as ordering things from the homeowner's account or opening the garage door to enter and steal from the home. Fourth, AI speakers who inadvertently misheard what users say as wake-up words record it and send it to the cloud. This could be an unintentional recording attack because the user did not attempt to communicate with the speaker (Chung et al., 2017).

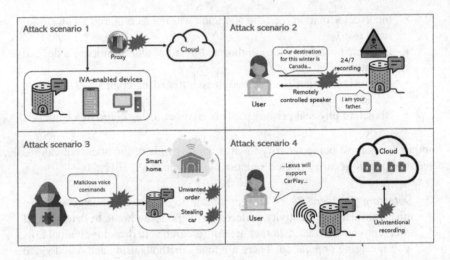

FIGURE 6.1 Four threat scenarios in an intelligent virtual assistant (IVA) ecosystem. (Adapted from Chung et al., 2017.)

O'Brolchain and Gordijn (2019) outlined recommendations to protect five perspectives on privacy in a SH as they pertain to vulnerable inhabitants:

1. *Informational* (e.g., controlling personal information),
2. *Physical* (e.g., solitude, modesty etc.),
3. *Associational* (e.g., medical context intimate experiences such as suffering and recovery from an illness),
4. *Proprietary* (e.g., anything related to a person's DNA), and
5. *Decisional* privacy (e.g., taking medication in the home).

Some of the recommendations to increase privacy include:

- Permitting residents to set privacy settings,
- Obtaining electronic consent, making sure the consent is understood by the inhabitant,
- Anonymizing data, and
- Securing medical data.

Researchers have also attempted many ways to safeguard personal information privacy for SH services. Wang et al. created a taint propagation analysis model to analyze how software-defined networking uses sensitive information and investigates suspicious security vulnerabilities. Traditional taint analysis is a method to check which variables can be modified/contaminated by user/attacker input. The taint propagation model uses a weighted spanning tree analysis scheme to track data. This method is focused on checking for exploits and suspicious behavior (Wang et al., 2016). Another solution for privacy management in the SH system is searchable encryption. Liu, Chen, and Huang (2019) proposed such a system – a searchable information retrieval of smart home data.

Typically, sensitive data is safeguarded with authentication methods. Authentication into the SH is a dominant theme among researchers as privacy is a challenging quality requirement to integrate into a SH system. Research in this area is at an early stage; thus, there are many recently proposed options. One authentication process proposed is access control mechanisms based on symmetric key cryptography. This process provides secure access to IoT home devices and stakeholders (Braeken et al., 2016). Other authentication schemes were proposed by Alshahrani and Traore (2019) and Raniyal et al. (2018). Alshahrani and Traore (2019) proposed and validated design of an authentication scheme with dynamic/temporary identities and cumulative Keyed-hash chain using various known attacks (i.e., replay, eavesdropping, impersonation, man-in-the-middle, session key guessing). Raniyal et al. (2018) proposed a two passphrase-protected device-to-device mutual authentication scheme.

In addition to data, another weak point/vulnerability in a SH are the many smart appliances. Every added device adds a new way to access your home. In other words, remote access of devices such as smart appliances is highly desirable for a SH system but come with a security risk. A more secure two-factor anonymous authentication scheme was proposed in Shuai (2019). Others proposed encryption schemes for lightweight remote user authentication into the SH environment (Naoui, Elhdhili & Saidane, 2019). The goal of these authentication schemes is to know who is accessing the SH devices.

Along those same lines, an intrusion detection system (IDS) was proposed for nefarious hackers. The idea is that the IDS is shared between the devices on the home network. This distributed IDS uses aggregated data from the devices that is analyzed by an expert system using an anomaly-based data analysis (Gajewski et al., 2019). Other researchers provided architectures or frameworks focusing on security and privacy in an SH design:

- SHSec-based architecture is a design to secure and manage the SH network. This is a middleware to ensure the interoperability of SH devices. It also includes threat prevention and mitigation of network security attacks (Sharma et al., 2019).
- An architecture to be able to control the home system the same way set-top boxes are used to stream multimedia was presented by Batalla and Gonciarz (2019). The home system is isolated from the Internet enabling secure communication from the homeowner to the SH.
- A Bluetooth digital, keyless door lock is proposed with an informal architecture (Dabhade et al., 2017). Bluetooth technology is more cost-efficient than radio frequency (RFID) and near-field communication (NFC) solutions. The system includes user detection, verification, notification, request processing, and emergency actions.
- A SH security framework providing an integrity system to prevent security attacks and promotes data integrity using self-signing and access control (Kang, Moon & Park, 2017).

- A lightweight authorization scheme when heterogeneous SH IoT devices are integrated in an already-existing untrusted cloud SH framework (Chifor et al., 2018).
- An architecture using blockchain to store device transactions (Singh et al., 2019). The SH system is able to detect denial-of-service (DoS) and distributed denial-of-service (DDoS) attacks. The architecture consists for four main components: *SH* (i.e., IoT devices), *blockchain* (i.e., a distributed ledger), *cloud computing* (i.e., manages client requests), and a *service layer* (i.e., interacts with the service providers and SH users).

A risk analysis of an SH automation system showed the human factor (e.g., connected devices exposing user privacy, information registry exposing routines) and software components of the system (e.g., application programming interfaces (APIs) and mobile apps) were classified as the highest risk (Jacobsson, Boldt & Carlsson, 2016). The results indicate these risks can be minimized if a more general model of security and privacy is included in the design phase of a SH with the following steps:

1. Identify and categorize SH personal data in transit,
2. Perform analysis and describe the main risks to privacy and security,
3. Identify and implement mitigating measures to reduce risk, and
4. Develop a strategy for SH information management.

Diagnostic testing devices could reveal vulnerabilities to improve the trustworthiness of the SH systems. When testing an IVA system, it is important to understand where data is generated and where it flows. For example, the malfunctions that occur during use by operating IVA-related services directly. Through the process of direct operation, we understand how the overall ecosystem of IVA is constructed. It also roughly selects potential attack points out of the entire ecosystem where cyberattacks are likely to take place. Potential attack points can be hardware level, software level, network level, cloud level, and so on. Empirical experiments and fuzz testing (i.e., software testing that involves invalid, unexpected, or random data as inputs) are also effective to show vulnerabilities at each possible attack point (Yu et al., 2020). Through experimentation, it distinguishes a traditional attack from a completely new type of attack. This can help us improve the reliability of IVAs in the future.

6.2.3 Physical Security

Data privacy and protection is clearly an important aspect to an SH but will not prevent a traditional home invasion. Thus, SHs just like a traditional home still need physical security – but it can be smarter than the average motion sensor that alerts the owner with the captured video of a squirrel running across your patio. For example, the traditional door chime and motion sensor to secure the physical security in a home can be transformed with IoT. Maybe we can determine who exactly is at the door with a video doorbell camera, for example. When someone presses the doorbell, the app on the smartphone alerts you to check who visited the house remotely.

There are real-time monitor services with high-definition cameras and will dispatch security personnel urgently if there is physical intrusion as well. Surantha and Wicaksono (2018) proposed a human detection system with 89% accuracy.

In summary, an SH is convenient because everything in the house can be controlled with a single device such as with mobile and AI speakers, but the related physical security market is naturally growing due to the high risk of hacking (Zeng, Mare & Roesner, 2017).

6.3 SMART HOME PRODUCTS/TOOLS/APPLICATION

SH products are those that focus on improving an inhabitant's comfort and automates some of their everyday tasks. The following sections will cover SH product examples followed by more specific considerations for SH product design.

6.3.1 SMART HOME PRODUCTS EXAMPLES

Plants are a lovely addition to a home – however, require continual care. A smart botanical garden with sensors to monitor the soil humidity, illumination, and temperature is an IoT device for the time-challenged. A smart garden with an emotion-aware feature that adjusts its growth and flowering according to the user's sleep schedule was described in Chen et al. (2017). Further plans for this system are to integrate security and privacy into the system.

Lee et al. (2017) introduced the concept of social connectedness (e.g., sense of belonging) and its integration with SH devices. Their research showed the effectiveness of SH devices when social connectedness is considered in the design. The typical SH IoT device is to facilitate the efficiency of the home residents' day-to-day tasks. Given that 28% of all households are single person (www.census.gov), many people who live alone would also benefit from devices that provide social support. Experimental study results showed that social connectedness with SH devices improved the perceived social support for users.

A home song recommender based on emotion analysis of texts collected during outdoor activities is proposed in Kang & Seo (2019). When the user's phone coordinated were more than the predefined distance from their home, texts were collected. The server continually evaluated the emotion of the texts and played the matching music upon the user's home arrival.

An application using a smart wall socket was developed to control the household appliances remotely using a mobile application (Phangbertha et al., 2019). There is a feature to turn devices on/off manually or by a timer. An electrical current monitoring feature is also under development.

Some unique IoT healthcare home products are also available. One product was equipped for a person challenged with dementia. A system was designed to detect and record home activities that were incomplete due to a person's condition (e.g., leaving the front door open; Demir et al., 2017). The system will remind/notify the resident with dementia as well as the caregiver of hazards with photographs.

Another product describing an integration with an SH dashboard includes notification for disease outbreaks using sentiment analysis on data mined and extracted from Twitter (Almazidy, Althani & Mohammed, 2016). And finally, a unique IoT health product was an ambient-assisted living SH product to assist people with special needs to age in their home called E-care@home (Alirezaie et al., 2017). For example, sensors are placed all around the home to detect activities such as eating and cooking.

6.3.2 SMART HOME PRODUCTS DESIGN CONSIDERATIONS

The hot topics for SH product design include improving specific development challenges, creating more products with a healthcare focus, and improving overall safety of SH products. Following are design recommendations for each of those topics (Alaa et al., 2017):

- Challenges: correct usage and monitoring power consumption,
- Healthcare: following medical guidelines and creating more products to assist the elderly and
- Safety: easing network and fault management

6.3.2.1 User Acceptance/User Experience

User acceptance is a highly researched topic. A survey was administered to analyze and understand the acceptance model of SH services (Park et al., 2018). The acceptance is based on attitude towards SH services, perceived usefulness, and ease of use. The results showed that the perceived usefulness has the most influence on attitude – where the usefulness depends on compatibility between the user's traditional devices and the new service. Another survey regarding user acceptance resulted in the understanding that users want to be able to access their SH services while outside of the home (Yang, Lee & Zo, 2018). This indicates that security, privacy, and trust are the vital factors of the provided SH services.

A comparative study with three of the most suitable SH control and management tools was conducted (Caivano et al., 2018). They chose the tools based on the following criteria: license and price, flexibility for users to add and modify devices in the SH, ease of controlling the device, tool extensibility (e.g., ease of integration with other home systems), technical support, ease of interconnections with other smart devices in the SH, interoperability with web services (e.g., file sharing, e-mail, weather apps, etc.). These criteria provide guidelines for SH management tool design. The top three tools were Atooma, IFITT, and Tasker. In general, all of the SH management tools evaluated need to address a better user experience.

6.3.2.2 Data Transfer

Within the SH, sending data from one sensor to another is a major part of the real-time monitoring process. Having this transfer occur in real time is the challenge. When comparing two methods, WebSocket and polling for data transfer, it was determined that the WebSocket methods were more effective (Soewito, Gunawan & Kusuma, 2019).

6.3.2.3 System Integration

Smirek, Zimmermann, and Beigl (2016) performed a comparative analysis with two platforms that address the user interfaces and interoperability issues holding up progress for an SH design effective for people with disabilities. The evaluated Eclipse Smart Home (ESH) and Universal Remote Console (URC) – they found these two systems complement each other and recommend the integration of these platforms.

6.3.2.4 Artificial Intelligence

Another literature review analyzed the work being done to integrate AI into SH applications (Zaidan & Zaidan, 2018). There is no doubt that AI integration into a SH is beneficial. These authors outlined many motivations related to system customization and efficiency, orchestration between appliances, energy efficiency, health benefits for the users, etc.

6.4 ACTIVITY/BEHAVIOR PATTERNS

Activity recognition is an important yet challenging research area for SHs. This technology is important for security (e.g., detect anomalies), comfort (e.g., heat/lighting settings) reasons, power consumption/reduction, and e-Health (e.g., rehabilitation, chronic disease management, and elderly monitoring). It requires robust, accurate, and efficient data analysis. It is complex because there may be multiple residents each with their own varying behavior. There are many strategies for sensing/analyzing/determining activities of daily living (ADLs) to identify a particular resident:

- Identify SH residents using behavioral patterns to distinguish the residents using a novel bag of sensors (BoF) approach that considers sensor events during a specific time frame (Lesani, Ghazvini & Amirkhani, 2019). A BoF approach considers the frequency of sensor events.
- Another approach to recognize complex activities of residents uses a formal concept analysis (FCA) strategy to recognize multiresident activities in the SH (Hao, Bouzouane & Gaboury, 2017, 2018). This approach uses sequential pattern mining and an incremental lattice search to determine the activity based on fewer sensor events.
- Activity tracking through an individual's behavioral patterns as well as changes in their regular activity – using simple sensors such as motion and door sensors. Using the Discovery Method for Varying Patterns (DMVP) sequences of sensor events occurring together and frequently allow activity tracking (Raeiszadeh, Tahayori & Visconti, 2019).
- The Complex Activity Recognition using Emerging patterns and Random forest (CARER) system focused on detecting more complex activities such as when multiple residents perform a simple activity in the same area at the same time or the living space is small (i.e., elder care; Malazi & Davari, 2018). In this approach, the sensors are put in place, data is extracted to

build a model, sequencing/segmenting of events is considered given their recency and sensor correlation, and then the segments are analyzed.

- Activity focus related to entrance and exit to a room using data from a passive infrared sensor (PIR; e.g., physical presence) and a hall effect sensor (e.g., door open/close; Skocir et al., 2016). This is useful to conserve heat/air conditioning/lighting systems as well as Ambient Assisted Living (AAL) applications.
- Optimizing human activity recognition is achieved by deep learning modeling techniques to learn high-level features from raw sensor data (Chen et al., 2018).
- An activity recognition system was designed through leveraging ontology to address activity diversity and Markov logic network (MLN) to address activity dynamics and uncertainty. In other words, the system to augment the ontology-based activity recognition with probabilistic reasoning (Gayathri, Easwarakumar & Elias, 2018).
- A web-based SH wellness monitoring system (WITS) to track/detect abnormal resident activity (Yao, et al., 2018). The system uses IoT middleware to learn from ambient sensor data. This system also enables specialists to diagnose wellness in real time.

ULD is a critical feature of a SH system. To address user privacy, device/tag inconvenience, and fault tolerance/accuracy, the ULD system exploits the prevalence of sensors in the home as well as a context broker (e.g., receives information from sensors) and a fuzzy-based decision-maker (e.g., evaluates the data from the broker; Ahvar et al., 2016b). In a literature survey of ULD methods discovered, there is not yet a definitive indoor ULD approach (e.g., outside uses GPS; Ahvar et al., 2016a). Accuracy and cost requirements need improvement. The recommendation that future designs should take advantage of more flexible, accurate, and ubiquitous ULD methods.

Developers also need data to test the robustness and scalability of their SH applications. Data was collected from a wrist wearable device to detect room localization accelerometer measurements. The data can help researchers when calibrating processes and methods for indoor localization systems in SHs (McConville et al., 2019).

6.5 POWER EFFICIENCY

Power is another predominant topic with cost reduction (energy efficiency), comfort, and safety as subthemes. Cost reduction can come from estimating energy consumption. Popa et al. (2019) presented a modular platform that creates user awareness from an advanced neural network model from the aggregated stored user data.

Home Energy Management System (HEMS) focuses on energy efficiency while maintaining user comfort. HEMs reduced consumption by 5.15% and comfort by 42.3% (Matsui, 2017). One of the most concerning issues with power is fires and shocks. A remotely controlled intelligent power outlet system can react fast and avoid overconsumption, fires, and shocks (Fernandez-Carames, 2015).

6.5.1 MANAGING ACTIVITY AND DAILY CONSUMPTION

Power consumption is also a common research area. In a field study with participants managing their energy tariff daily to match their consumption level, participants showed to be ready to use a Tariff Agent to manage their energy tariffs (Alan et al., 2016). The findings imply that this activity is a way to sustain engagement with the system.

Understanding the usage patterns for more effective resource planning of a smart grid is critical. An effective SH approach, focusing on cost reduction, is to enable the homeowner to plan appliance usage. Gupta, Jha, and Nagar (2016) created an approach to increase convenience to resource planning and cost reduction with their proposed approach to a smart grid that is a reliable technology to control and monitor electricity and information to generate, transmit, and distribute electricity.

6.6 SYSTEMS DESIGN

In SH design, researchers focused on requirements, both functional and nonfunction, and architecture. Functional requirements are those that describe the services a system provides and how the system reacts to input (Laplante, 2014). Nonfunctional requirements are slightly more challenging as they are difficult to define and verify (e.g., safety, reliability, usability, interoperability). Software architecture involves the structure, interaction, and organization of the system's components and subcomponents to form a system (Laplante, 2007). There are many architectural styles, each representing important properties of the system.

6.6.1 ARCHITECTURE

A SH high-level architecture is made up of three main components (Li et al, 2018):

1. *System Master*: Servers, communication networks, workstations, and interactions with applications, websites.
2. *Family Intelligent Interactive Terminal*: Interface into the system components (appliances, security, etc.)
3. *Smart Electrical Equipment*: Appliances, security, electrical systems data collection, etc.

6.6.2 REQUIREMENTS

The major requirements to provide an SH platform with which to build more robust SH applications were provided along with suggested architecture styles (Hui, Sherratt & Sanchez, 2017).

1. *Heterogeneity (Server Centralized Architecture and Service-Oriented Architecture)*: Enabling information exchange between things connected to the network.
2. *Self-Configurable (Wireless Sensor Network Architecture – WSN)*: Enabling the addition or removal of connected devices in the SH network.

3. *Extensibility (WSN)*: Enabling the extension or configuration update of the connected devices.
4. *Context Awareness (Open Services Gateway Initiative Architecture – OSGi)*: The capability to detect location or environmental changes.
5. *Usability (Brain–Computer Interface – BCI)*: Ease of use and easy to learn the SH system.
6. *Security and Privacy Protection (WSN)*: Level of protection against unauthorized use and malicious attacks.
7. *Intelligence (Data-Information-Knowledge-Wisdom – DIKW)*: Enabling human activity prediction.

Interoperability is also an essential requirement of an SH system as integrating heterogeneous devices to enable communication. Santofimia et al. (2018) proposed a virtual protocol that supports communication among different technology. Puustjarvi and Puustjarvi (2015) went a step further and created SH ontology to facilitate semantic interoperability in an SH. Semantic interoperability enables data exchange between devices with unambiguous shared meaning. They used the principles of Linked Data (e.g., a best practice for providing a data infrastructure to share data across the web) to integrate data sources from external data sources as well – all to increase the usability of the system within a SH.

For device management, which can facilitate device interoperability, the requirements and design of a fog-assisted cloud architecture were proposed. Fog computing environment provides a means to share network services between cloud data centers and end users. A technique called Resource Management Technique for Smart homes (ROUTER) for fog-enabled cloud computing environments was proposed. This technique optimizes the response time, network bandwidth, energy consumption, and latency of the devices (Gill, Garraghan & Buyya, 2019).

IoT device management not only needs to facilitate interoperability between devices, but also needs to be efficient. A prevalent research area is energy efficiency. Jo and Yoon (2018) create three IoT platform models to work together to allow the IoT devices to cooperate as well as reduce network congestion and conserve energy.

1. *Intelligence Awareness Target as a Service (IAT)*: Learns situational awareness of the data generated by sensors.
2. *Intelligence Energy Efficiency as a Service (IE^{2}S)*: Processes the data collected by IAT to learn energy usage patterns.
3. *Intelligence Service TAS (IST)*: Provides control and management of the services.

Efficient device communication and management are vital requirements of an SH. For SH device communication (i.e., efficient initialization, synchronization, and data transmission), a model based on visible light communication (VLC) was proposed. This model (Tiwari, Sewaiwar & Chung, 2017) uses light-emitting diodes to transmit and photodetectors as receivers. Another presented a novel cuckoo search algorithm to optimize VLC (Sun et al., 2017).

Asghari and Cheriet (2018) proposed an SH network that promotes energy efficiency and the lifetime of the SH network. This framework also considered the optimal measures to manage sensor node traffic. Along those same lines to provide a more energy-efficient offloading algorithm for home automation applications was proposed (Zhang et al., 2018).

SH products don't always take the user's emotions in mind – leading to a lack of adoption. Curumsing et al. (2019) developed an emotion-oriented requirements engineering approach to model to identify, model, and evaluation of emotional goals. They applied this technique to develop a SH platform for the elderly that included passive monitoring and emergency assistance. SH technical accident prevention (e.g., a gas, smoke, or water leak). A subsystem model was another product design (Teslyuk et al., 2018).

6.6.3 SIMULATIONS/MODELING

Context information (i.e., temperature, humidity, sound) and user modeling and adaptation are pervasive part of the SH. Thus, effective user modeling is critical. The system must take into account users' cognitive level and physical disabilities. A user model that focuses on extracting a user's cognitive activity rather than movement is more effective (Vlachostergiou et al, 2016).

SH simulation systems helped to improve the SH design to meet the needs of the residents with minimal investment. Using agent modeling, SHs can adapt to the owner's needs (Vasilateanu & Bernovici, 2018). Testing and determining the most effective SH configuration of the IoT sensors and actuators will meet the personalized needs of the residents.

6.7 CONSTRUCTING AN IoT HOME DEVICE

Often, life can be challenging. We are constantly working, taking care of ourselves, family, and learning new things. However, sometimes everyday tasks can get in the way of what we need to do, but they don't always have too. This section will explain two SH devices that are easy to build and can help someone really grasp the concept of IoT: a smart garden and a smart trash can.

6.7.1 SMART GARDEN

Watering plants can sometimes be a burden and is easy to forget. A smart garden uses a simple program to monitor and care for a plant. While it seems like a minor convenience, it really is so much more than that. The smart garden is a symbol of the infinite possibilities that the IoT provides.

The smart garden, Figure 6.2, is a device that demonstrates how a computer program can automatically nurture and grow a plant. The raspberry pi, a credit card-sized computer, is used in the device to allow a software application to collect and act on the plant data. Using a sensor, the application measures the plant levels and distributes the correct amount of water to the plant based off of the data – a true IoT device. The application is able to do this infinitely as long as water is accessible.

FIGURE 6.2 Smart garden.

The smart garden is a lot more efficient in keeping a plant alive than the average person – as the plant is never over or under watered. The purpose is to keep the plant alive so the plant owner can do more important things.

Specifically, the smart garden sensors measure a plant's temperature, soil moisture, amount of sunlight, and air temperature. All of this data is aggregated and analyzed by the application program which decides when and how much to water the plant. A smart garden kit can be purchased (ex. SwitchDoc Labs (https://shop.switch-doc.com/). The kit comes with most of what you need to build this device. The only things that need to be purchased separately are the raspberry pi, a computer monitor/keyboard, a power supply, and a plant.

Once you have assembled the smart garden according to the kit, the application has hard-coded range of the plant levels: moisture, temperature, etc. Once the moisture is too low, the sensor, that is inserted into the soil, sends the data to the raspberry pi, which sends a signal to the water pump to begin watering the plant. The pump stops watering once the sensor detects enough moisture.

While the smart garden is an efficient and easy way to take care of plants, it does have a few flaws. First of all, it's not a practical size. It requires many wires and tubes, which makes it difficult for the user to enjoy the beauty of the plant. Second, an entire system can only sustain one plant. It seems like quite a great deal of effort for a single plant. However, this experiment is not about its practicality, it is about the system itself which clearly can be improved. The smart garden has limitless potential.

This project is an easy and fantastic example of IoT possibilities and a great way to learn about IoT. It shows that an everyday task can be made effortless. It is a small project with potential to be updated into something that can make a tremendous impact on a home.

6.7.2 SMART TRASHCAN

A smart trashcan is one of the possible components of a SH implementation as waste management. This could also be a vital function of a smart city concept. The collection of trash has a fixed schedule, but as the generation of the trash is dynamic in nature, it may be possible that one trashcan fills up earlier than another. This creates problems not only by the bad odor but also from the perspective of environmental factors. By building a smart trashcan, the concerned people can be notified well in advance about which can pick up first.

6.7.2.1 Concept

The best way to detect the level of filled trash in a trashcan is by using an ultrasonic sensor attached to the lid as shown below in Figure 6.3.

Ultrasonic sensor mounted on the lid will face downward and send trigger waves towards the bottom of the trashcan. When the can is filled up to a certain level, waves will bounce back as echo which will be captured back by the sensor. The time difference between the sending of trigger signal and reception of echo signal provides a reasonable measure of the level of the garbage in the trashcan.

This system can be simulated with the following components:

- *IBM Bluemix IOT Service*: IBM offers its inhouse IoT messaging service based on MQTT protocol. It is available as part of the IBM Bluemix catalog. A sign-up for IBM Bluemix trial account will be needed.

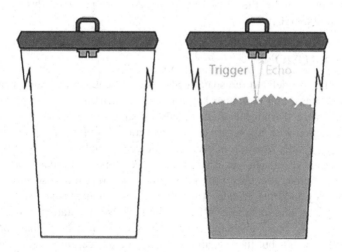

FIGURE 6.3 Trashcan with ultrasonic sensor attached.

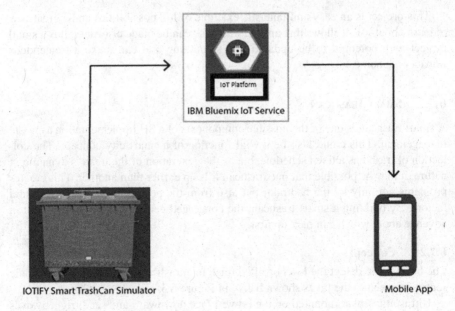

FIGURE 6.4   Architecture for the smart trashcan.

- *IOTIFY Smart Trashcan Simulator*: This is the complete hardware emulation of a trashcan, available under IOTIFY. You will need to sign up for IOTIFY subscription to access this.
- *Android Mobile (Optional)*: In order to monitor this virtual trashcan via mobile, one needs an Android mobile phone. The architecture of the system is depicted in Figure 6.4.

The technical details of the virtual trashcan setup are provided through the link: https://bit.ly/3ax36GU.

6.8   CONCLUSION

In this chapter, we defined the span of SH research, categorized and identified the SH research areas, and presented key findings in each category. Several improvement areas emerged from the studies analyzed such as improved user experience, human interaction with SH systems, and integration of wellness/healthcare devices. On the security side, research is needed in detecting multiperson activity patterns and keeping ahead of new security threats. Therefore, threat-detecting technologies need to improve to reduce false positives and to prevent new security risks and increase privacy and trust. Besides, future SH designs should take advantage of more flexible, accurate, and ubiquitous ULD methods – also facilitating improvement in SH security. In summary, given the research presented, great strides have been made in SH development – but there is always room to enhance, improve, and innovate, especially with security and privacy.

FURTHER READING

Ahvar, E., Daneshgar-Moghaddam, N., Ortiz, A., Lee, G., Crespi, N., "On analyzing user location discovery methods in smart homes: a taxonomy and survey," *Journal of Network and Computer Applications*, vol. 76, 2016a, pp. 75–86.

Ahvar, E., Lee, G., Han, S., Crespi, N., Khan, I., "Sensor network-based and user-friendly user location discovery for future smart homes," *Sensors*, vol. 16, no. 969, 2016b, pp. 1–17.

Alaa, M., Zaidan, A.A., Zaidan, B.B., Talal, M., Kiah, M., "A review of smart home applications based on Internet of Things," *Journal of Network and Computer Applications*, vol. 97, 2017, pp. 48–65.

Alan, A., Costanza, E., Ramhurn, S., Fischer, J., Rodden, T., Jennings, N., "Tariff agent: interacting with a future smart energy system at home," *ACM Transactions on Computer-Human Interaction*, vol. 23, no. 4, 2016, pp. 1–28.

Alirezaie, M., Renoux, J., Kockemann, U., Kristoffersson, A., Karlsson, L., Blomqvist, E., Tsiftes, N., Voigt, T., Loutfi, A., "An ontology-based context-aware system for smart homes:E-care@home," *Sensors*, vol. 17, 2017, pp. 1–23.

Almazidy, A., Althani, H., Mohammed, M., "Towards a disease outbreak notification framework using twitter mining for smart home dashboards," *Procedia Computer Science*, vol. 82, 2016, pp. 132–134.

Alshahrani, M., Traore, I., "Secure mutual authentication and automated access control for IoT smart home using cumulative Keyed-Hash chain," *Journal of Information Security and Applications*, vol. 45, 2019, pp. 156–175.

Asghari, V., Cheriet, M., "Energy and connectivity aware resource optimization of nodes traffic distribution in smart homes networks," *Future Generation Computer Systems*, vol. 88, 2018, pp. 559–570.

Batalla, J. M., Vasilakos, A., Gajewski, M., "Secure smart homes: opportunities and challenges," *ACM Computing Surveys*, vol. 50, no. 5, 2017, pp. 1–32.

Batalla, J., Gonciarz, F., "Deployment of smart home management system at the edge: mechanisms and protocols," *Neural Computing and Applications*, vol. 31, 2019, pp. 1301–1315.

Braeken, A., Porambage, P., Stojmenovic, M., Lambrinos, L., "eDAAAS: efficient distributed anonymous authentication and access in smart homes," *International Journal of Distributed Sensor Networks*, vol. 12, no. 12, 2016, pp. 1–11.

Caivano, D., Fogli, D., Lanzilotti, R., Piccinno, A., Cassano, F., "Supporting end users to control their smart home: design implications from a literature review and an empirical investigation," *The Journal of Systems and Software*, vol. 144, 2018, pp. 295–313.

Chen, G., Wang, A., Zhao, S., Liu, L., Chang, C., "Latent feature learning for activity recognition using simple sensors in smart homes," *Multimedia Tools Applications*, vol. 77, 2018, pp. 15201–15219.

Chen, M., Yang, J., Zhu, X., Wang, X., Liu, M., Song, J., "Smart Home 2.0: innovative smart home system powered by botanical IoT and emotion detection," *Mobile Network Applications*, vol. 22, 2017, pp. 1159–1169.

Chifor, B., Bica, I., Patriciu, V., Pop, F., "A security authorization scheme for smart home Internet of Things devices," *Future Generation Computer Systems*, vol. 86, 2018, pp. 740–749.

Chung, H., Iorga, M., Voas, J., Lee, S., "Alexa, can i trust you?," *Computer*, vol. 50, 2017, pp. 100–104.

Curumsing, M., Fernando, N., Abdelrazek, M., Vasa, R., Mouzakis, K., Grundy, J., "Emotion-oriented requirements engineering: a case study in developing a smart home system for the ederly," *Journal of Systems and Software*, vol. 147, 2019, pp. 215–229.

Dabhade, J., Javare, A., Ghayal, T., Shelar, A., Gupta, A., "Smart door lock system: improving home security using bluetooth technology," *International Journal of Computer Applications*, vol. 160, no. 8, 2017, pp. 19–22.

Dahmen, J., Cook, D., Wang, X., Hongler, W., "Smart secure homes: a survey of smart home technologies that sense, assess, and respond to security threats," *Journal of Reliable Intelligence Environment*, vol. 3, 2017, pp. 83–98.

Demir, E., Koseoglu, E., Sokullu, R., Seker, B., "Smart home assistant for ambient assisted living of elderly people with dementia," *Procedia Computer Science*, vol. 113, 2017, pp. 609–614.

Fernandez-Carames, T., "An intelligent power outlet system for the smart home of the Internet of Things," *International Journal of Distributed Sensor Networks*, vol. 11, 2015, pp. 1–11.

Gajewski, M., Batalla, J., Mastorakis, G., Mavromoustakis, C., "A distributed IDS architecture model for smart home systems," *Cluster Computer*, vol. 22, 2019, pp. 1739–1749.

Gayathri, K., Easwarakumar, K., Elias, S., "Probabilistic ontology based activity recognition in smart homes using Markov logic network," *Knowledge-Based Systems*, vol. 121, 2017, pp. 173–184.

Gill, S., Garraghan, P., Buyya, R., "ROUTER: fog enabled cloud based intelligent resource management approach for smart home IoT devices," *The Journal of Systems and Software*, vol. 154, 2019, pp. 125–138.

Golvilkar, B., Gosavi, A., Pawar, B., Mohite, A., Magar, P., Shete, P., "Novel security of home using smart home management system," *International Journal of Advanced Research in Computer Science*, vol. 9, no. 2, 2018, pp. 337–339.

Gupta, R., Jha, D., Nagar, S., "Agent based smart home energy management system," *CSI Transactions on ICT*, vol. 4, 2016, pp. 103–110.

Hao, J., Bouzouane, A., Gaboury, S., "Complex behavioral pattern mining in non-intrusive sensor-based smart homes using an intelligent activity inference engine," *Journal of Reliable Intelligence Environment*, vol. 3., 2017, pp. 99–116.

Hao, J., Bouzouane, A., Gaboury, S., "Recognizing multi-resident activities in non-intrusive sensor-based smart homes by formal concept analysis," *Neurocomputing*, vol. 318, 2018, pp. 75–89.

Heartfield, R, Loukas, G., Budimir, S., Bezemskij, A., Fontaine, J., Filippoupolitis, A., Roesch, E., "A taxonomy of cyber-physical threats and impact in the smart home," *Computers & Security*, vol. 78, 2018, pp. 398–428.

Hui, T., Sherratt, S., Sanchez, D., "Major requirements for building smart homes in smart cities based on Internet of Things technologies," *Future Generation Computer Systems*, vol. 76, 2017, pp. 358–369.

Jacobsson, A., Boldt, M., Carlsson, B., "A risk analysis of a smart home automation system," *Future Generation Computer Systems*, vol. 56, 2016, pp. 719–733.

Jo, H., Yoon, Y., "Intelligent smart home energy efficiency model using artificial TensorFlow engine," *Human-centric Computing and Information Sciences*, vol. 8, no. 9, 2018, pp. 1–18.

Kang, D., Seo, S., "Personalized smart home audio system with automatic music selection based on emotion," *Multimedia Tools and Applications*, vol. 78, 2019, pp. 3267–3276.

Kang, W., Moon, S., Park, J., "An enhanced security framework for home appliances in smart home," *Human-Centric Computing and Information Sciences*, vol. 7, no. 6, 2017, pp. 1–12.

Laplante, P., *Requirements Engineering for Software and Systems*. CRC Press, Boca Raton, FL, 2014.

Laplante, P., *What Every Engineer Should Know about Software Engineering*. CRC Press, Boca Raton, FL, 2007.

Lee, B., Kwon, O., Lee, I., Kim, J., "Companionship with smart home devices: the impact of social connectedness and interaction types on perceived social support and companionship in smart homes," *Computers in Human Behavior*, vol. 75, 2017, pp. 922–934.

Lesani, F., Ghazvini, F., Amirkhani, H., "Smart home resident identification based on behavior patterns using ambient sensors," *Personal and Ubiquitous Computing*, vol. 6, 2019, pp. 1–12.

Li, M., Gu, W., Chen, W., He, Y., Wu, Y., Zhang, Y., "Smart home: architecture, technologies and systems," *Procedia Computer Science*, vol. 131, 2018, pp. 393–400.

Liu, H., Chen, G., Huang, Y., "Smart hardware hybrid secure searchable encryption in cloud with IoT privacy management for smart home system," *Cluster Computer*, vol. 22, 2019, pp. 1125–1135.

Malazi, H., Davari, M., "Combining emerging patterns with random forest for complex activity recognition in smart homes," *Applied Intelligence*, vol. 48, 2018, pp. 315–330.

Marr, B., "The 5 biggest smart home trends in 2020," Forbes Magazine, January 13, 2020. https://www.forbes.com/sites/bernardmarr/2020/01/13/the-5-biggest-smart-home-trends-in-2020/#7724bf96389b, retrieved 10/6/2020.

Matsui, K., "An information provision system to promote energy conservation and maintain indoor comfort in smart homes using sensed data by IoT sensors," *Future Generation Computer Systems*, vol. 82, 2017, pp. 388–394.

McConville, R., Byrne, D., Craddock, I., Piechocki, R., Pope, J., Santos-Rodriguez, R., "A dataset for room level indoor localization using a smart home in a box," *Data in Brief*, vol. 22, 2019, pp. 1044–1051.

Naoui, S., Elhdhili, M., Saidane, L., "Lightweight and secure passwords based smart home authentication protocol: LSP-SHAP," *Journal of Network and Systems Management*, vol. 27, 2019, pp. 1020–1042.

O'Brolchain, F., Gordijn, B., "Privacy challenges in smart homes for people with dementia and people with intellectual disabilities," *Ethics and Information Technology*, vol. 21, 2019, pp. 253–265.

Park, E., Kim, S., Kim, Y., Kwon, S. J., "Smart home services as the next mainstream of ICT industry: determinants of the adoption of smart home services," *Universal Access in the Information Society*, vol. 17, 2018, pp. 175–190.

Phangbertha, L., Fitri, A., Parnamasari, I., Muliono, Y., "Smart socket for electricity control in home environment," *Procedia Computer Science*, vol. 157, 2019, pp. 465–472.

Popa, D., Pop, F., Serbanescu, C., Castiglione, A., "Deep learning model for home automation and energy reduction in a smart home environment platform," *Neural Computing and Applications*, vol. 31, 2019, pp. 1317–1337.

Puustjarvi, J., Puustjarvi, L., "The role of smart data in smart home: health monitoring case," *Procedia Computer Science*, Vol. 69, 2015, pp. 143–151.

Raeiszadeh, M., Tahayori, H., Visconti, A., "Discovering varying patterns of normal and interleaved ADLs in smart homes," *Applied Intelligence*, vol. 49, 2019, pp. 4175–4188.

Raniyal, M., Woungang, I., Dhurandher, S., Ahmed, S., "Passphrase protected device-to-device mutual authentication schemes for smart homes," *Security Privacy*, vol. 1, 2018, p. e42.

Santofimia, M., Villa, D., Acena, O., Toro, X., Trapero, C., Villanueva, F., Lopez, J., "Enabling smart behavior through automatic service composition for Internet of Things-based smart homes," *International Journal of Distributed Sensor Networks*, vol. 14, no. 8, 2018, pp. 1–20.

Sharma, D., Baldeo, A., Phillip, C., "Raspberry Pi based smart home deployment in the smart grid," *International Journal of Computer Applications*, vol. 119, no. 4, 2015, pp. 6–10.

Sharma, P.K., Park, J.H., Jeong, Y., Park, J.H., "SHSec: SDN based secure smart home network architecture for Internet of Things," *Mobile Networks and Applications*, vol. 24, 2019, pp. 913–923.

Shuai, M., Yu, N., Wang, H., Xiong, L., "Anonymous authentication scheme for smart home environment with provable security," *Computers and Security*, vol. 86, 2019, pp. 132–146.

Singh, S., Ra, I., Meng, W., Kaur, M., Cho, G., "SH-BlockCC: a secure and efficient Internet of things smart home architecture based on cloud computing and blockchain technology," *International Journal of Distributed Sensor Networks*, vol. 15, no. 4, 2019, pp. 1–18.

Skocir, P., Krivic, P., Tomeljak, M., Kusek, M., Jezic, G., "Activity detection in smart home environment," *Procedia Computer Science*, vol. 96, 2016, pp. 672–681.

Smirek, L., Zimmermann, G., Beigl, M., "Just a smart home or your smart home – a framework for personalized user interfaces based on eclipse smart home and universal remote console," *Procedia Computer Science*, vol. 98, 2016, pp. 107–116.

Soewito, B., Gunawan, F. E., Kusuma, I. G. P., "Websocket to support real time smart home applications," *Procedia Computer Science*, vol. 157, 2019, pp. 560–566.

Sun, G., Liu, Y., Yang, M., Wang, A., Liang, S., Zhang, Y., "Coverage optimization of VLC in smart homes based on improved cuckoo search algorithm," *Computer Networks*, vol. 116, 2017, pp. 63–78.

Surantha, N., Wicaksono, W., "Design of smart home security system using object recognition and PIR sensor," *Procedia Computer Science*, vol. 135, 2018, pp. 465–472.

Teslyuk, V., Beregovskyi, V., Denysyuk, P., Teslyuk, T., Lozynskyi, A., "Development and implementation of the technical accident prevention subsystem for the smart home system," *International Journal of Intelligent Systems and Applications*, vol. 1, 2018, pp. 1–8.

Tiwari, S.V., Sewaiwar, A., Chung, Y., "Smart home multi-device bidirectional visible light communication," *Photon Network Communications*, vol. 33, 2017, pp. 53–59.

Vasilateanu, A., Bernovici, B., "Lightweight smart home simulation system for home monitoring using software agents," *Procedia Computer Science*, vol. 138, 2018, pp. 153–160.

Vlachostergiou, A., Stratogiannis, G., Caridakis, G., Siolas, G., Mylonas, P., "User adaptive and context aware smart home using pervasive and semantic technologies," *Journal of Electrical and Computer Engineering*, vol. 2016, 2016, pp. 1–20.

Wang, P., Chao, K., Lo, C., Lin, W., Lin, H., Chao, W., "Using malware or software-defined networking-based smart home security management through a taint checking approach," *International Journal of Distributed Sensor Networks*, vol. 12, no. 8, 2016, pp. 1–23.

Yang, H., Lee, H., Zo, H., "User acceptance of smart home services: an extension of the theory of planned behavior," *Industrial Management & Data Systems*, vol. 117, no. 1, 2018, pp. 68–89.

Yao, L., Sheng, Q., Benatallah, B., Dustdar, S., Wang, X., Shemshadi, A., Kanhere, S., "WITS: an IoT-endowed computational framework for activity recognition in personalized smart homes," *Computing*, vol. 100, 2018, pp. 369–385.

Yu, M., Zhuge, J., Cao, M., Shi, Z., Jiang, L., "A survey of security vulnerability analysis, discovery, detection, and mitigation on IoT devices," *Future Internet*, vol. 12, no. 2, 2020, p. 27.

Zaidan, A., Zaidan, B., "A review on intelligent process for smart home applications based on IoT: coherent taxonomy, motivation, open challenges, and recommendations," *Artificial Intelligence Review*, vol. 53, 2018, pp. 141–165.

Zeng, E., Mare, S., Roesner, F., "End user security & privacy concerns with smart homes", *Thirteenth Symposium on Usable Privacy and Security*, USENIX Association, Santa Clara, CA, 2017.

Zhang, J., Zhou, Z., Li, S., Gan, L., Zhang, X., Qi, L., Xu, X., Dou, W., "Hybrid computation off loading for smart home automation in mobile cloud computing," *Personal and Ubiquitos Computing*, vol. 22, 2018, pp. 121–134.

7 IoT in Education[1]

7.1 INTRODUCTION

According to analyst firm Gartner, 8.4 billion 'things' were connected to the Internet in 2017; excluding the laptops, computers, tablets, and mobile phones (Gartner, 2017). In addition, an estimate predicts this number to increase to 64 billion Internet of Things (IoT) devices worldwide by 2025 (Petrov, 2020). Regardless of the exact number of devices, spending in this market is expected to increase substantially. The International Data Corporation (IDC) is forecasting worldwide IoT spending to reach $1.1 trillion in 2023 (IDC, 2019).

As we have been discussing throughout the book, IoT applications are already being leveraged in diverse domains such as the medical services field, smart retail, customer service, smart homes, environmental monitoring, and industrial internet. Now, because of their ubiquitous nature, schools and academic institutions are looking to incorporate IoT in educational activities to benefit students, instructors, and the entire educational system. IoT applications are being proposed to address a diverse range of modes, objectives, subjects, and perceptions in the education sector. While some of the scenarios for IoT in education may be apparent, others are not so obvious.

For example, because IoT devices can monitor things, IoT can be used to monitor student attendance and in-class activities as proposed in Alotaibi (2015) and Jiang (2016). However, monitoring can also be applied in other ways. Some studies focus on monitoring the students and objects in online education and online laboratories' settings (e.g., Fernandez et al., 2015; Lamri et al., 2014; Bisták, 2014; Yin et al., 2012; Bin, 2012; Srivastava & Yammiyavar, 2016; Shi et al., 2010). IoT is also being deployed for more pragmatic scenarios in education. For example, in Valpreda & Zonda (2016), an IoT-based system was proposed to increase children's knowledge about agricultural food production and consumption. Other studies on deploying IoT to educate students with special needs exist (e.g., Sula et al. 2013, 2014) for children with autism spectrum disorder.

Educators in the USA (GSMA, 2012), Japan (Fuse, Ozawa & Miura, 2012), and the UK (UNESCO, 2014) already have been applying IoT technologies in their pedagogical processes. Also, several academic institutions (e.g., University of San Francisco (Nie, 2013)) have incorporated IoT technology to enhance campus safety. The information technology and networking giant Cisco has launched several IoT projects for the education sector through engaging schools and students in a smart academic forum (Cisco, 2013). Despite the several studies available on the inclusion of IoT into the education domain, there is still a lack of consolidated and coherent views on this subject. On the other hand, although beneficial, the adoption of IoT technologies in education also poses some new challenges to their realization.

---

[1] Excerpts of this chapter are from Kassab, DeFranco and Laplante (2020).

DOI: 10.1201/9781003027799-7

With the increased use of IoT in the education domain, it is of utmost importance to study how this technology with its distinguished systems functions, such as sensing and decision-making, can support and challenge the pedagogical processes for all interrelated actors (faculty, students, and staff) as well as all involved assets (e.g., libraries, classrooms, laboratories).

Although there have been several contributions on the inclusion of IoT into the education domain, there is still lack of consolidated and coherent views on this subject. Hence, the discussions in this chapter aim to close the gap of knowledge and embark on mapping out the published studies available. In addition, the results of a systematic literature review focused on the benefits and the challenges faced in education in integrating IoT into the curriculum and educational environments will be presented. Different mapping views of the extracted studies are provided as well as a summary of the already implemented tools and a list of gap research questions yet to be investigated. We also present a tool for an adaptive learning experience in response to a remote learner's emotions. This chapter will answer the following questions:

Q1. What are the benefits of the adopted scenarios of IoT in education?

Q1.1. Which education level(s) are addressed with the discussed educational scenarios?

Q1.2. Which education subject(s) are addressed with the discussed educational scenarios?

Q1.3. Which perspective(s) (e.g., instructor, student, staff) are addressed with the discussed educational scenarios?

Q1.4. Which learning principles are addressed with the discussed educational scenarios?

Q1.5. Which education settings (e.g., face to face, online) are addressed with the discussed educational scenarios?

Q2. What are the challenges of incorporating IoT in education?

To answer the above questions, we used the systematic literature review (SLR) method (Keele, 2007) to identify, extract, evaluate, and synthesize the available articles on the symbiosis of IoT in education. The results show that we identified a significant variety of positive contributions for IoT technologies on education. In particular, we identified three dimensions to classify emerging scenarios of IoT in education and to discuss the benefits: delivery mode, perception, and learning principles (Figure 7.1). While the focus of this chapter is on answering the above questions, we also raise a set of gap research questions that we discuss in Section 7.5, and we also provide a demonstration of a tool we developed to provide an adaptive learning experience in response to a remote learner's emotions in Section 7.4.

The presentation in this chapter will be mostly in the format of reporting on the results from the SLR we followed. Nevertheless, the description of the SLR process we conducted is beyond scope of this chapter. If the reader is interested in learning more about the methodology we followed in the preparation of the content of

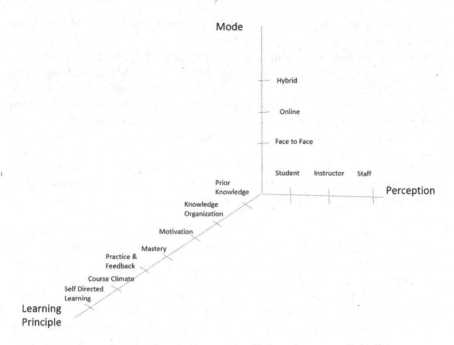

FIGURE 7.1 The three-dimensional scheme for Internet of Things (IoT) in education.

this chapter, we invite him/her to read our published article (Kassab, DeFranco & Laplante, 2020).

7.2 WHAT ARE THE BENEFITS OF THE ADOPTED SCENARIOS OF IoT IN EDUCATION?

7.2.1 APPLYING IoT IN DIFFERENT EDUCATION LEVELS AND SUBJECTS

The literature suggests that IoT can be incorporated at all education levels. We provide a mapping of the primary selected studies into the educational levels in: https://bit.ly/2XCZf5K. We also find that IoT incorporation is not limited to particular education subjects. The literature presents scenarios related to "Computer, Electrical and Electronics Engineering" education, "Science", "language skills" education, "Physical education", "Physiological education", and "Business education".

7.2.2 PERCEPTION

Perception is a way of regarding, understanding, or interacting with something. More specifically, in the context of this chapter, perception refers to a stakeholder role that calls on the IoT system to deliver one of its services in education settings. A stakeholder has a goal concerning the system – one that can be satisfied by its operation. We identified from the extracted scenarios three different primary stakeholders, each with perception when interacting with the IoT technologies when deployed in the education field: instructor, student, and staff.

As for the *instructor*, IoT can help to manage attendance of a class and availability of required equipment/devices for each student (Alotaibi, 2015; Jiang, 2016; Borges et al., 2011; Qi & Shen, 2011; Gul et al., 2017). "Installing RFID reader at the entrance of school gate, library, cafeteria, dormitory and teaching building, and other places to identify students' RFID electronic tags, it can obtain the students' activities trajectory" (Jiang, 2016). Besides, with IoT, an instructor may initiate and manage class session with voice/facial/gesture commands (Fuse, Ozawa & Miura, 2012; He & Zhuang, 2016; Zhu, 2016), communicate with remote students at different locations (Sarıta,s, 2015; Uskov et al., 2016; Shirehjini, 2016), and collect immediate feedback from students in terms of interests in an activity or lesson and sensor data.

Analytics could also be run on the sensor data to evaluate behavior, performance, interest, and participation of each student and provide a summary to the instructor (Elyamany & AlKhairi, 2015; Haiyan & Chang 2012; Ueda & Ikeda, 2016; Richert et al., 2016). IoTs can help the instructor to confirm the identity of students (Wang, 2015; Zhu, 2016; Tan et al., 2014); it can also help the instructor to identify and help students with special needs (Sula et al., 2014; Lenz et al., 2016).

From a *student* perspective, IoT will help to communicate with classmates (local or remote; Yin et al., 2012), share project data, discuss and annotate learning materials in a real time (Bin, 2012), and access the learning resources remotely (e.g., remote labs; Fernandez et al., 2015; Lamri et al., 2014; Bisták, 2014, De la Torre, Sánchez & Dormido, 2016; García et al., 2013; Chunxia, 2015; Kane et al., 2013; Tunc et al., 2015). Also, IoT could provide support to students with adapted learning resources by integrating content that is based on location, time, date, student-to-student interaction, knowledge level, etc. (Sula et al., 2013, 2014; Möller, Haas & Vakilzadian, 2013; Peña-Ríos et al., 2012; Jeong, Kim & Chong, 2015; Murphy et al., 2015; Lenz et al., 2016; Chen et al., 2011; Ma & Li, 2013; Gómez, Huete & Hernandez, 2016).

From a *staff* perspective, IoT will play a designated role in elements such as tracking students (Pena-Rios et al., 2012; Xue et al., 2011). For example, an IoT scenario is reported in Wang (2014) on monitoring and maintaining psychological health for students. Another reported contribution is the potential assistance for staff members in managing and tracking fixed and portable academic resources (Han, 2011; Caţă, 2015). "Using a noise sensor, one classroom can communicate automatically to a neighbor classroom and inform them if the noise level exceeds a certain level. A warning message could be displayed on the LCD screen in the noisy room" (Caţă, 2015). For the public portable equipment (e.g., portable projectors, laboratory equipment, sports equipment), these can be marked with a tag to be tracked by the Radio Frequency Identification (RFID) technology (Bin, 2012). The collected data from tracking portable equipment can be further utilized to automatically calculate patterns and trends and find inefficiencies (Li & Li, 2014). IoT can also assist staff members in managing events (e.g., registration events (Tan et al., 2014), sports events (Ma & Li, 2013)) and in managing the general safety and security (Fuse, Ozawa & Miura, 2012; Elyamany & AlKhairi, 2015; Caţă, 2015). In addition, it can also play a role in institutional energy management (Cisco, 2013; Caţă, 2015).

7.2.3 LEARNING PRINCIPLES

In Ambrose et al. (2010), the authors listed seven principles that underlie the foundations for effective learning. These principles are student prior knowledge, knowledge organization, motivation, mastery, practice and feedback, course climate, and
self-directed learning. These principles are distilled from research from a variety of
disciplines (CMU, 2017). Figure 7.2 shows the distribution of the extracted scenarios
we collected from our search in literature along with the seven principles.

Our findings indicate that IoT technologies will make a positive impact on each of
these seven principles. While there is a slight bias in the extracted scenarios towards
utilizing IoT technologies to improve the "course climate" in particular, every learning principle can be enhanced with IoT. For example, in Sula et al. (2014) the authors
propose a smart assistive environment system that uses a Heuristic Diagnostic
Teaching (HDT) process where the intention is to identify each student's learning
abilities in math as well as their creativity traits. Their proposed system uses a computer, sensors, RFID tag reader, and a SmartBox device in order support learning for
students with Autism spectrum disorder (ASD)by providing a personalized "practice
and feedback" on a case-by-case basis. Other studies that discuss utilizing IoT to
provide personalized learning experience include Sula et al. (2013, 2014); Möller,
Haas & Vakilzadian (2013); Peña-Ríos et al. (2012); Meda, Kumar & Parupalli
(2014); Jeong, Kim & Chong (2015); Murphy et al. (2015); Lenz et al. (2016); Chen
et al. (2011); Wang (2010); Ma & Li (2013); Gómez, Huete & Hernandez (2016).

In Wan (2016), improving student's "motivation" principle is addressed through a
proposed scenario aiming at bridging the communication between teachers and students using IoT. In Elyamany and AlKhairi (2015), the authors proposed an innovative system based on IoT to analyze the impact of several parameters of the physical

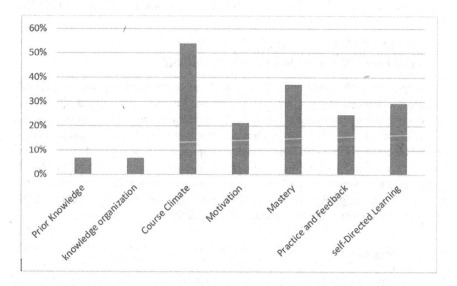

FIGURE 7.2 Distribution of the extracted scenarios from our search in literature along with
the seven principles.

environment in a classroom on students' focus, where the term "focus" refers to the students' subjective feeling of their ability to concentrate on a lecture at a given moment. The primary goal was to identify those parameters that significantly affect students' focus in the course climate. Studies with a similar goal include Uzelac, Gligoric & Krco (2015); Ueda & Ikeda (2016); Richert et al. (2016). In Antle, Matkin & Warren, (2016) the Story of Things (SoT) system is proposed to enable children to learn the story behind every object they touch on a typical day. Inspired by Living Media and the IoT, "the goal is to change children's awareness through hands-on interaction with the world they live in. A back-of-the-hand display is activated by stick-on finger sensors when a child touches an object. They can tap the display to select from several stories stored in a crowdsourced database about that object (e.g., the materials it was made from; the processes used to make it; how it impacts their body; how it will be disposed of; environmental or social rights challenges associated with the object; and how they can take positive action)". This information is overlaid on the world through an augmented-reality contact lens to enhance the "knowledge organization".

Another example of utilizing IoT for "knowledge organization" principle is reported in Gómez et al. (2013) in which the authors discuss a project utilizing IoT to improve a child's attitude towards food via learning about food consumption and production and ways to reduce waste on a-long-term basis. A complete mapping of the selected studies into the seven learning principles is reported in https://bit.ly/2Yj3gts.

7.2.4 DELIVERY MODE

Education can be delivered in one of three broad-based modes: face to face, remote, or hybrid. The findings from the SLR we conducted indicate that the selected studies were almost equally distributed between the face-to-face setting and the online setting. IoT has been discussed in all of these three modes: https://bit.ly/2Xu0zYM.

7.3 WHAT ARE THE CHALLENGES OF INCORPORATING IoT IN EDUCATION?

When specifying the functionality for IoT education applications, attention is often focused on concerns such as fitness of purpose, Big Data, interoperability, and so on. Conventional requirement elicitations techniques such as Quality Function Deployment (QFD), Joint Application Development (JAD), and domain analysis among others (Laplante, 2017) are usually adequate for these types of requirements. But in IoT applications for education, specific quality requirements are probably of greater concern. We identified three major qualities from the extracted scenarios that may pose a challenge for IoT in education: security, scalability, and humanization. We explore these three qualities further in this section.

7.3.1 SECURITY

Security requirements have always been a crucial aspect of education. Given the increased communication and complexity of IoT technology, there is an increase in security-related concerns (Georgescu & Popescu, 2015). Out of the selected studies,

20% of the papers discussed "security/privacy" concerns, making it the most discussed quality (Alotaibi, 2015; Lamri et al., 2014; Lei et al., 2016; Samoilă, Ursuțiu & Jinga, 2016; Brady et al., 2015; Qi & Shen, 2011; Heng, Yi & Zhong, 2011; Ueda & Ikeda, 2016; Cață, 2015; Gul et al., 2017; García et al., 2013; Hui & Haiyan, 2011; Putjorn, Ang & Farzin, 2015; De la Guía et al. 2016; Murphy et al., 2015; Georgescu & Popescu, 2015; Kane et al., 2013; Tan et al. 2014). Two of these studies, namely Putjorn, Ang & Farzin, (2015) and Murphy et al. (2015) discussed particularly the challenge of child privacy when using IoT for education. It has become increasingly clear that educational systems are vulnerable to cyberattacks, and the number of attacks is predicted to increase (Weber, 2010). Students can easily stage cyberattacks on their institutions, or schools/universities could be prevented from functioning as intended. "Cascade failures may appear, caused by the interconnectivity of a large number of devices, difficult to be simultaneously protected over the air transmission, with all the related problems" (Georgescu & Popescu, 2015).

Education, and particularly higher education, is often identified as having a large number of reported data breaches, and at first look, the Privacy Rights Clearinghouse (PRC) database appears to confirm this view. In the United States, there were 727 reported breaches in educational institutions between the years 2005 and 2014 (Grama, 2014). This number is the second highest among seven sectors that were investigated (the first is healthcare). About 7% of all academic institutions in the United States have had a least one breach. From 2005 to 2014, 66% of academic institutions listed in the PRC experienced only one reported breach. However, about one-third of institutions with breaches have had more than one. Six percent of the listed institutions have experienced five or more reported breaches. Hacking/malware where an outside party accessed records via direct entry, malware, or spyware was the largest proportion of the reported breaches at 36%. "Many of the devices used in a provisioned, specialized IoT will collect various data whether that surveillance is known or not" (Laplante et al., 2015). But why are these data being collected? Who owns the data? And where does the data go? These are questions that need to be answered by the legal profession, government entities that oversee education, and educational standards groups. For example, in July 2000, the Higher Education Information Security Council (HEISC) was established. The HEISC provides coordination and support about information security governance, compliance, and data protection and privacy to higher education institutions. To help a better understanding of the nuance of information security issues in higher education, members of the HEISC drilled down into the topic of information security and identified their top three strategic information security issues for 2016 (Grama & Vogel, 2016). "Planning for and implementing next-generation security technologies" with increasing concerns of the IoT is one of the three strategic issues.

7.3.2 Scalability

By embedding sensors into front field environments as well as terminal devices, an IoT network can collect rich sensor data that reflect the real-time environment conditions of the front field and the events/activities that are going on. Advanced data mining technologies can be applied to explore in-depth business insights from these data. Since the data is collected in the granularity of elementary event level in a 7×24 mode, the data volume is very high and the data access pattern also differs considerably from traditional business data. This has motivated a new generation of data management solutions, e.g., NoSql database, map-reduce distributed computing framework, etc. (Zhang, 2016).

The IoT in the education domain is not an exception. Incorporating IoT in education will generate a large volume of data. Hence, the need for analyzing and treating these data to capture information and trends emerges. The scalability concern is addressed in seven papers out of the selected primary studies. In Jagtap et al. (2016), the authors address the issue of scalability in the context of providing personalized content to the students based on analyzing a large volume of collected student data and activity. They propose a design for a "social recommender" system that is based on Hadoop and its parallel computing platform.

In Mehmood et al. (2017), the authors discuss the scalability concern when designing a learning management system. Other selected studies that discussed the scalability issue are Lamri et al. (2014); Sandu, Costache & Balan (2015); Pei et al. (2013); Jurkovicová et al., (2015); and Alvarez, Silva & Correia, (2016). With scalability, concerns regarding the discussion on the cost of the IoT technology in education becomes also significant. The main question that may arise is whether in the long 20 runs, IoT devices and Big Data analysis will increase the already-existing divide into a two-class learning system: those who can disburse these technologies and those who cannot! At the same time, if going to school should be affordable for everyone, how will schools be able to buy and service these devices? The financial concern of IoT in education is discussed in four selected articles: Maleko, Hamilton & D'Souza (2012); Putjorn, Ang & Farzin (2015); Georgescu & Popescu (2015); and Pruet et al., (2015).

7.3.3 HUMANIZATION

There are questions on the moral role that IoT may play in human lives, particularly in respect to personal control. Applications in the IoT involve more than computers interacting with other computers. Fundamentally, the success of the IoT will depend less on how far the technologies are connected and more on the humanization of the technologies that are connected (Tech, 2015). IoT technology may reduce people's autonomy, shift them towards particular habits, and then shifting power to corporations focused on financial gain. For the education system, this effectively means that the controlling agents are the organizations that control the tools used by the academic professionals but not the academic professionals themselves (Gubbi et al., 2013).

Dehumanization of humans in interacting with machines is a valid concern, and it is discussed in two selected papers: Murphy et al. (2015) and Lenz et al. (2016). Many studies indicate that face-to-face interaction between students will not only benefit a child's social skills but also positively contribute towards the character building. The issue that may arise from increased IoT technologies in education is the partial loss of the social aspect of going to school. Conversely, using IoT in virtual learning environments can be of special support to students of special needs (e.g., dyslexic and dyscalculic needs, for example Lenz et al. (2016)).

IoT can offer students with special needs the opportunity to randomly often repeat experiments without major damage to property or cost. Thus, the students with special needs could feel reduced levels of frustration and feel less self-conscience since they could have more time to repeat an experiment. In addition, a prejudice-free performance evaluation may be possible with the anonymization (Lenz et al., 2016). For dyslexic and dyscalculic students; e.g., it is likely that anonymization will be advantageous for them as possible functions because of IoT (e.g., autocorrect) will improve any bias with a student's score because teachers will not know if they are assessing students with a learning challenge or not.

7.4   MONITORING EMOTIONAL STATE OF ONLINE LEARNER: A TOOL

There are no adequate empirically proven strategies to address the presence of emotions in education. Incorporating IoT technologies combined with the power of Big Data analytics to support the detection of (and reacting) to the learner's emotional state during online learning experience can positively improve the learners' motivation and satisfaction with the course climate, which may decrease the drop-out rate in an on-line program. The captured data could also lead to insights that can be made functional for the well-being of the students and improve the provided "feedback". For example, a simple webcam combined with an already-existing emotion detection artificial programming interface (API) can be utilized for an affordable and nonintrusive emotional-based elearning to detect the facial expressions of the remote learner. The webcam can also be used to capture the learners' eye-tracking (detecting a learner's gaze location). Companies that provide webcam eye-tracking services include GazeHawk and EyeTrackShop. Fitness monitoring devices, such as Fitbit, can also be utilized to take the input of the heartbeat.

With this background in mind, we developed a Learning Management System (LMS) that supports capturing and reacting to remote learner's emotions in real time. The LMS was developed using Python 3.7. The system client APIs are JQuery, material for page UI, and Websocket for communicating with a web server. The system server APIs are as follows: Tornado framework for web service, OpenCV, Keras CNN model, and TensorFlow API for real-time face detection Towards Designing Smart Learning Environments with IoT and emotion/gender classification.

The integrated system captures the user's expressions based on the gender against the online content and time stamp. It detects seven emotional expressions: angry, disgust, fear, happy, sad, surprise, and neutral. Once a learner logs into the website, the system triggers a pop-up message notifying that the webcam will be turned on. After acknowledging the message and while navigating through the online materials (Figure 7.3), the system captures the facial expressions of a learner. The tool captures the user's webcam frame data with websocket addressed once in every 100 ms.

The system evaluates the expressions once every 10 seconds as per customized parameters which can be modified by a course administrator. The system detects whether user's expressions match with the customized parameters while navigating particular content (e.g., a percentage of emotional expressions (sad, happy, etc.) within any 10 seconds time frame). If there is a match, then the system prompts a

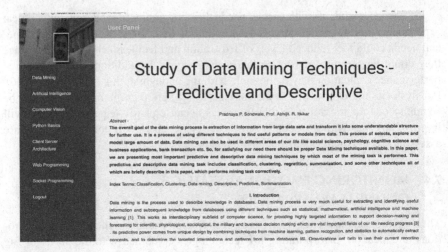

FIGURE 7.3   A screenshot from the tool at run time while a remote learner navigates through online content.

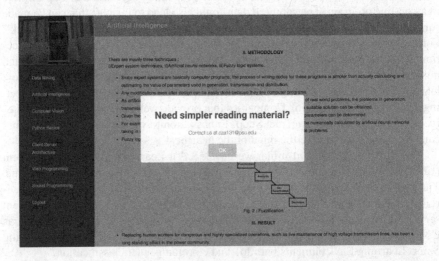

FIGURE 7.4   A screenshot from the tool at run time. The tool detects a particular percentage of emotional expressions and prompts the learner with supplemental learning materials corresponding to the content.

learner with the option of supplemental learning materials customized according to the learner's state and the particular content (Figure 7.4). In addition, an "aggregator" component sends periodically aggregated data on the learner's profile and the entire online class state to the course instructor.

The captured data can serve the well-being of the individual student and for the entire class. While the initial phase of the project successfully integrated a facial emotional expression API, we are currently working on the next phase of the tool to integrate eye-gaze tracking and heartbeats monitoring APIs. These two additional

TABLE 7.1

Online Learner's State Use Case Construct Model

Model Element	Realization
1. Sensor	Webcam, fitbit
2. Snapshot (time)	Once in every 100 ms
3. Cluster	Set of (2) proximity sensors per online learner
4. Aggregator	Determine learning state of the learner
5. Weight	Room layout dependent
6. Communication channel	Compliant network of sensors or clusters or aggregator, wired (Internet) to *e*Utility
7. eUtility	Remote monitoring software
8. Decision	Personalized content

types of detentions will allow us to position more precisely the state of a remote learner simultaneously while reading the course content, participating in the online designated discussions forums or writing the course assignments. A simple construct for monitoring the online learner state use case scenarios using an IoT as described in this section is shown in Table 7.1.

7.5 DISCUSSION

Looking at all the extracted studies on IoT in education, we find that 21% of the overall studies provided actual implementation for IoT systems which were tested in actual educational settings. We provide through a link (https://bit.ly/2XD23ji) a summary of these implemented systems along with information on how these were tested in practice and the testing results.

We further analyzed these systems and extracted the below list of the most common implemented scenarios which aim to integrate IoT in education:

- Smart environment for supporting learning and improving the quality of education
- Elearning and education management
- Virtual learning, teaching, and management
- Remote laboratories and distant elaboratories
- Human motion capture
- Ubiquitous learning, distance eTeaching, and eLearning
- Augmented reality elearning
- Wireless robotic educational platform
- Cybersecurity laboratory
- Attendance system based on IoT
- Psychological health education service based on IoT

On the other hand, this literature review demonstrated that there are significant research gaps related to IoT education and its derivatives (such as cyber-physical systems). These gaps are driven by the large number and ever-evolving technologies, widely diverse application domains, the many applicable standards, and the cross-disciplinary nature of IoT education. Some of the gap research questions that we identified are as follows:

7.5.1 WHAT ARE THE RIGHT DEVICES AND PROCESSING COMPONENTS FOR IoT PEDAGOGY?

There are many IoT technologies including various sensors, processing elements, and cloud-based services, and it would be impossible to represent a comprehensive set in any academic laboratory or classroom. Research is needed to understand a sufficient and cost-effective subset of technologies for the development and delivery of the IoT curricula and experimental laboratories.

In our SLR, we assembled a list of the IoT tools discussed within educational settings from those studies that provided implementation for IoT systems in educational settings. We provide the list through the link: (https://bit.ly/2XD23ji). As we mentioned, 64% of the extracted studies proposed a system architectural solution for IoT in educational settings but only 21% of the overall papers provided actual implementation. All of the implemented systems were tested for usability or effectiveness with quantitative and/or qualitative methodologies. We provide a summary of these implemented systems through the link (https://bit.ly/2XD23ji) along with how these systems were integrated into educational settings and tested. We also provide a summary of the testing results for these systems.

7.5.2 HOW ARE INSTRUCTORS, STAFF, AND STUDENTS GOING TO CONNECT AND USE THE IoT NETWORK FOR TEACHING AND LEARNING?

When we investigated the subset of the tools extracted from the prime studies and which are proposed to be used by students, instructors, or staff to interact with the IoT systems, smartphones/mobile devices were the most commonly mentioned (42% of the overall extracted; Figure 7.5), followed by sensors/RFID technologies (27% of the overall extracted scenarios for each). Selecting the right technology mix for an IoT curriculum is important, but students and instructors will need to access that technology from within and outside the classroom and laboratories. There are significant security, safety (e.g., in cyber-physical systems), and privacy issues surrounding IoT systems as we discussed in the previous section. There are also issues involving low power consumption (e.g., for battery-powered IoT devices), high power requirements (for large cyber-physical systems), and bandwidth utilization limits, especially for programs and courses with many students. Therefore, further research is needed towards defining a safe and efficient networking structure for IoT program delivery.

7.5.3 WHAT CAN EDUCATIONAL INSTITUTES DO TO MITIGATE THESE COMPLEXITIES?

IoT systems and the related hardware and software are highly complex. Many applicable standards apply to IoT systems and their components and these need to be taught in the curriculum and educational systems will need to comply with these standards.

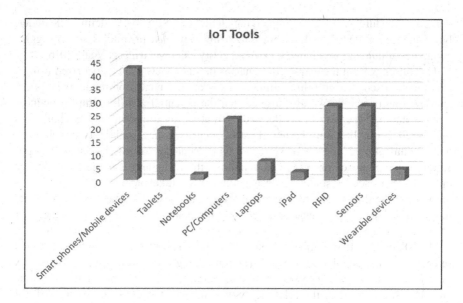

FIGURE 7.5 Tools used to interact with Internet of Things (IoT) systems in educational settings.

International standardization bodies' industrial organizations publish hundreds of standards that can apply to IoT systems, both domain-specific and cross-domain. For example, the IEEE (and its association and industrial partners) supports nearly 100 standards applicable to IoT (https://standards.ieee.org/initiatives/iot/). Therefore, any IoT-based curricula must address the identification of appropriate standards, harmonization of differences, and standards blending.

7.6 CONCLUSION

The advances in sensors, nano-electronics, smart objects, cloud computing, Big Data, and communication on a wide scale will make innovation continuous in IoT, and it will clout a great number of domains. The education domain is not an exception. While IoT education is a new conceptual paradigm and it is still in its starting phase, IoT is set to transform the education domain in many ways shortly.

This systematic review presented in this chapter is a starting point for constructing an in-depth understanding of the impact of IoT devices and applications in the education domain. This review demonstrated that the curiosity with utilizing IoT applications has well-reached education discipline and that it has been accompanied by increasing research publications.

As the realm of IoT protocols and standards expands, adding these standards to the education system could also increase the complexity for all its members. While mitigating this complexity is open for research, the US National Institute of Standards and Technologies (NIST) Special Publication SP 800-183, "Networks of Things," which was presented earlier in this book, should form the basic building blocks of IoTs and a framework for researching and developing IoT scenarios for education

domain and addressing the accompanying challenges. SP 800-183 defines the underlying science for the IoT by describing the technology's five primitives (sensor, aggregator, a communications channel, external entity, decision trigger; Voas, 2016).

Our findings from the extracted scenarios in this review were categorized using a
classification scheme with three dimensions related to education mode, perception,
and learning principle. We also argued that the potential of IoT to improve educational outcomes needs to be moderated with attention to the important challenges
uncovered in this literature review. That is, security issues, scalability of solutions,
and humanization of the delivery system. New scenarios for IoT in education applications to be proposed or built need to address these challenges while delivering an
optimal experience for students, teachers, and other stakeholders.

Also, while this review revealed that there is no lack of studies on the potential
usages of IoT on every education level, only three studies addressed the usages for
children who are 5 years or younger (Miglino et al., 2014; Murphy et al., 2015; Lei
et al., 2016). This low number suggests that more research needs to be conducted
in regard to this population, particularly concerning integration with the right mix
of IoT technologies while respecting the constraints due to the population's unique
characteristics. Besides, the relatively low numbers of studies addressing IoT for
"knowledge organization" and "prior knowledge", learning principles suggest that
more scenarios worth to be investigated for these principles.

FURTHER READING

Alotaibi, S. J., "Attendance system based on the Internet of Things for supporting blended
learning," *2015 World Congress on Internet Security (WorldCIS)*, Dublin, Ireland,
pp. 78–78, 2015.

Alvarez, I. B., Silva, N. S. A., Correia, L. S., "Cyber education: towards a pedagogical
and heuristic learning," *ACM SIGCAS Computers and Society*, vol. 45, no. 3, 2016,
pp. 185–192.

Ambrose, S.A., Bridges, M. W., DiPietro, M., Lovett, M. C., Norman, M. K., *How Learning
Works: Seven Research-Based Principles for Smart Teaching*. John Wiley & Sons,
Hoboken, NJ, 2010.

Antle, A. N., Matkin, B., Warren, J., "The story of things: awareness through Happenstance
interaction," *International Conference on Interaction Design and Children*, pp. 745–750,
Manchester, UK, 2016.

Bin, H., "The design and implementation of laboratory equipments management system
in university based on Internet of Things," *International Conference on Industrial
Control and Electronics Engineering*, Xi'an, China, pp. 1565–1567, 2012.

Bisták, P., "Mobile web application for group of remote laboratories using middleware,"
*IEEE 12th IEEE International Conference on Emerging eLearning Technologies and
Applications (ICETA)*, Stary Smokovec, Slovakia, pp. 51–56, 2014.

Borges, V., Sawant, R., Zarapkar, A., Azgaonkar, S., "Wireless automated monitoring sys-
tem for an educational institute using Learning Management System (MOODLE),"
International Conference of Soft Computing and Pattern Recognition (SoCPaR),
Dalian, China, pp. 231–236, 2011.

Brady, C., Weintrop, D., Gracey, K., Anton, G., Wilensky, U., "The CCL-Parallax programmable badge: learning with low-cost, communicative wearable computers," *Conference
on Information Technology Education*, Chicago, IL, pp. 139–144, 2015.

Caţă, M., "Smart university, a new concept in the Internet of Things," *Roedunet International Conference-Networking in Education and Research (roedunet ner)*, Craiova, Romania, pp. 195–197, 2015.

Chen, J., Man, H., Jin, Q., Huang, R., "Goal-driven navigation for learning activities based on process optimization," *International Conference on Internet of Things and 4th International Conference on Cyber, Physical and Social Computing*, Dalian, China, pp. 389–395, 2011.

Chunxia, J., "Laboratory management of the Internet based on the technology of Internet of Things," *AASRI International Conference on Industrial Electronics and Applications (IEA 2015)*, Auckland, 2015.

Cisco, "Education and the Internet of Everything – how ubiquitous connectedness can help transform pedagogy," 2013. https://bit.ly/2JvJqHB.

CMU, "Principles of Learning," 2017. https://www.cmu.edu/teaching/principles/learning.html.

De la Guía, E., Camacho, V. L., Orozco-Barbosa, L., Luján, V. M. B., Penichet, V. M., Pérez, M. L., "Introducing IoT and wearable technologies into task-based language learning for young children," *IEEE Transactions on Learning Technologies*, vol. 9, no. 4, 2016, pp. 366–378.

De la Torre, L., Sánchez, J. P., Dormido, S., "What remote labs can do for you," *Physics Today*, vol. 69, no. 4, 2016, pp. 48–53.

DeFranco, J., Kassab, M., Voas, J., "How do you create an Internet of Things workforce?," *IT Professional*, vol. 20, no. 4, 2018, pp. 8–12.

Elyamany, H. F., AlKhairi, A. H., "IoT-academia architecture: a profound approach," *IEEE/ACIS 16th International Conference on Software Engineering, Artificial Intelligence, Networking and Parallel/Distributed Computing (SNPD)*, Takamatsu, Japan, pp. 1–5, 2015.

Fernandez, G. C., Ruiz, E. S., Gil, M. C., Perez, F. M., "From RGB led laboratory to servo-motor control with websockets and IoT as educational tool," *Proceedings of 2015 12th International Conference on Remote Engineering and Virtual Instrumentation (REV)*, Bangkok, Thailand, pp. 32–36, 2015.

Fuse, M., Ozawa, S., Miura, S., "Role of the Internet for risk management at school," *International Conference on Information Technology Based Higher Education and Training (ITHET)*, Istanbul, Turkey, pp. 1–6, 2012.

García, M. L., Fernandez, G. C., Ruiz, E. S., Martín, A. P., Gil, M. C., "Rethinking remote laboratories: widgets and smart devices," *IEEE Frontiers in Education Conference (FIE)*, Oklahoma City, OK, pp. 782–788, 2013.

Gartner, "The Internet of Things (IoT) is a key enabling technology for digital businesses," 2017. http://www.gartner.com/technology/research/internet-of-things/.

Georgescu, M., Popescu, D., "How could Internet of Things change the E-learning environment," *The 11th International Scientific Conference eLearning and Software for Education*, Bucharest, 2015.

Gómez, J. E., Huete, J. F., Hernandez, V. L., "A contextualized system for supporting active learning," *IEEE Transactions on Learning Technologies*, vol. 9, no. 2, 2016, pp. 196–202.

Gómez, J., Huete, J. F., Hoyos, O., Perez, L., Grigori, D., "Interaction system based on Internet of Things as support for education," *Procedia Computer Science*, vol. 21, 2013, pp. 132–139.

Grama, J., "Just in time research: data breaches in higher education," EDUCAUSE, 2014.

Grama, J., Vogel, V., "The 2016 Top 3 strategic information security issues," 2016. https://bit.ly/1KCEH2j.

GSMA, "Mobile education in the United States," 2012. https://www.gsma.com/iot/mobile-education-in-the-united-states/.

Gubbi, J., Buyya, R., Marusic, S., Palaniswami, M., "Internet of Things (IoT): A vision, architectural elements, and future directions," *Future Generation Computer Systems*, vol. 29, no. 7, 2013, pp. 1645–1660.

Gul, S., Asif, M., Ahmad, S., Yasir, M., Majid, M., Malik, M., Arshad, S., "A survey on role of Internet of Things in education," *International Journal of Computer Science and Network Security*, vol. 17, no. 5, 2017, pp. 159–165.

Haiyan, H., Chang, S., "The design and implementation of ISIC-CDIO learning evaluation system based on Internet of Things," *World Automation Congress*, Puerto Vallarta, Mexico, pp. 1–4, 2012.

Han, W. 2011. "Research of Intelligent Campus System Based on IOT." In: Jin D., Lin S. (eds) *Advances in Multimedia, Software engineering and Computing*, vol. 1, pp. 165–169. Springer, Berlin.

He, B.-X., Zhuang, K.-J., "Research on the intelligent information system for the multimedia teaching equipment management," *International Conference on Information System and Artificial Intelligence (ISAI)*, Hong Kong, China, pp. 129–132, 2016.

Heng, Z., Yi, C. D., Zhong, L. J., "Study of classroom teaching aids system based on wearable computing and centralized sensor network technique," *International Conference on Internet of Things and 4th International Conference on Cyber, Physical and Social Computing*, Dalian, China, pp. 624–628, 2011.

Hui, Y., Haiyan, H., "Research and realization on CDIO teaching experimental system based on RFID technique of Internet of Things," *International Conference on Mechatronic Science, Electric Engineering and Computer (MEC)*, Jilin, China, pp. 841–844, 2011.

International Data Corporation, "Steady commercial and consumer adoption will drive Worldwide spending on the Internet of Things to $1.1 Trillion in 2023, according to a new ICD guide," June 13, 2019. https://www.idc.com/getdoc.jsp?containerId=prUS45197719, retrieved 10/18/2019.

Jagtap, A., Bodkhe, B., Gaikwad, B., Kalyana, S., "Homogenizing social networking with smart education by means of machine learning and Hadoop: a case study," *International Conference on Internet of Things and Applications (IOTA)*, Pune, India, pp. 85–90, 2016.

Jeong, K., Kim, H.-S., Chong, I., "Knowledge driven composition model for WoO based self-directed smart learning environment," *International Conference on Information Networking (ICOIN)*, Cambodia, pp. 537–540, 2015.

Jiang, Z., "Analysis of student activities trajectory and design of attendance management based on Internet of Things," *2016 International Conference on Audio, Language and Image Processing (ICALIP)*, Shanghai, China, pp. 600–603, 2016.

Jurkovicová, L., Cervenka, P., Hrivíková, T., Hlavaty, I., "E-learning in augmented reality' utilizing iBeacon technology," *Proceedings of the European Conference on E-Learning*, Hatfield, United Kingdom, pp. 170–178, 2015.

Kane, P., Duda, M., Farrell, S., Jeffords, J., Kimsey, T., Rucinski, A., Zhong, J. 2013. "Disruptive engineering training and education based on the Internet of Things," *IEEE International Conference on Teaching, Assessment and Learning for Engineering (TALE)*, Bali, Indonesia, pp. 632–636, 2013.

Kassab, M., DeFranco, J., Laplante, P., "A systematic literature review on Internet of Things in education: Benefits and challenges," *Journal of Computer Assisted Learning*, vol. 36, no. 2, 2020, pp. 115–127.

Keele, S., 2007. Guidelines for performing systematic literature reviews in software engineering (Tech. Rep.). Technical report, Ver. 2.3 EBSE Technical Report. EBSE.

Lamri, M., Akrouf, S., Boubetra, A., Merabet, A., Selmani, L., Boubetra, D., "From local teaching to distant teaching through IoT interoperability," *2014 International Conference on Interactive Mobile Communication Technologies and Learning (IMCL2014)*, Thessaloniki, Greece, pp. 107–110, 2014.

Laplante, P. A., Laplante, N., Voas, J., "Considerations for healthcare applications in the Internet of Things," *Reliability*, vol. 61, no. 4, 2015, pp. 8–9.

Laplante, P. A., *Requirements Engineering for Software and Systems*. Auerbach Publications, Boston, MA, 2017.

Lei, L., Dai, Q., Wang, M., Liu, Q., Xiao, M., "The research and implementation of engineering training system based on mobile Internet of Things," *IEEE International Conference on Consumer Electronics-China (ICCE-China)*, Nanchang, China, pp. 1–4, 2016.

Lenz, L., Pomp, A., Meisen, T., Jeschke, S., "How will the Internet of Things and Big Data analytics impact the education of learning-disabled students? A concept paper," *3rd MEC International Conference on Big Data and Smart City (ICBDSC)*, Muscat, Oman, pp. 1–7, 2016.

Li, J., Li, X., "Design of management system for teaching equipment based on the Internet of Things," *Contemporary Logistics*, vol. 15, 2014, p. 33.

Ma, G., Li, Y., "Application of IoT in Information Teaching of Ethnic Colleges," *Proceedings of the 2013 International Conference on Information, Business and Education Technology (ICIBET–2013)*, Beijing, 2013.

Maleko, M., Hamilton, M., D'Souza, D., "Novices' perceptions and experiences of a mobile social learning environment for learning of programming," *Proceedings of the 17th ACM Annual Conference on Innovation and Technology in Computer Science Education*, Haifa, Israel, pp. 285–290, 2012.

Meda, P., Kumar, M., Parupalli, R., "Mobile augmented reality application for Telugu language learning," *IEEE International Conference on MOOC, Innovation and Technology in Education (MITE)*, Patiala, India, pp. 183–186, 2014.

Mehmood, R., Alam, F., Albogami, N. N., Katib, I., Albeshri, A., Altowaijri, S. M., "UTiLearn: a personalised ubiquitous teaching and learning system for smart societies," *IEEE Access*, vol. 5, 2017, pp. 2615–2635.

Miglino, O., Di Fuccio, R., Di Ferdinando, A., Ricci, C., "BlockMagic, a hybrid educational environment based on RFID technology and Internet of Things concepts," *International Internet of Things Summit*, pp. 64–69, Rome, 2014.

Möller, D. P., Haas, R., Vakilzadian, H., "Ubiquitous learning: teaching modeling and simulation with technology," *Grand Challenges on Modeling and Simulation Conference*, p. 24, Toronto, 2013.

Murphy, F. E., Donovan, M., Cunningham, J., Jezequel, T., García, E., Jaeger, A., Popovici, E. M., "i4Toys: video technology in toys for improved access to play, entertainment, and education," *IEEE International Symposium on Technology and Society (ISTAS)*, Dublin, Ireland, pp. 1–6, 2015.

Nie, X., "Constructing smart campus based on the cloud computing platform and the Internet of Things," *2nd International Conference on Computer Science and Electronics Engineering*, Hangzhou, 2013.

Pei, X. L., Wang, X., Wang, Y. F., Li, M. K., "Internet of Things based education: definition, benefits, and challenges," *Applied Mechanics and Materials*, vol. 411, pp. 2947–2951, 2013.

Peña-Ríos, A., Callaghan, V., Gardner, M., Alhaddad, M. J., "Remote mixed reality collaborative laboratory activities: Learning activities within the InterReality Portal," *IEEE/WIC/ACM International Joint Conferences on Web Intelligence and Intelligent Agent Technology*, vol. 3, pp. 362–366, Macau, 2012.

Pena-Rios, A., Callaghan, V., Gardner, M., Alhaddad, M. J., "Towards the next generation of learning environments: an InterReality learning portal and model," *Eighth International Conference on Intelligent Environments*, pp. 267–274, Guanajuato Mexico, 2012.

Petrov, C., "47 Stunning Internet of Things Statistics 2020 [The rise of IoT]", Tech Jury, October 13, 2020. https://techjury.net/blog/internet-of-things-statistics/#gref, retrieved 10/18/2020.

Pruet, P., Ang, C. S., Farzin, D., Chaiwut, N., "Exploring the Internet of "Educational Things" (IoET) in rural underprivileged areas," *12th International Conference on Electrical Engineering/Electronics, Computer, Telecommunications and Information Technology (ECTI-CON)*, pp. 1–5, Hua Hin, 2015.

Putjorn, P., Ang, C. S., Farzin, D., "Learning IoT without the I-educational internet of things in a developing context," *Workshop on Do-It-Yourself Networking: An Interdisciplinary Approach*, pp. 11–13, Florence, 2015.

Qi, A.-Q., Shen, Y.-J., "The application of Internet of Things in teaching management system," *International Conference of Information Technology, Computer Engineering and Management Sciences*, vol. 2, pp. 239–241, Nanjing, 2011.

Richert, A., Shehadeh, M., Plumanns, L., Groß, K., Schuster, K., Jeschke, S., "Educating engineers for industry 4.0: virtual worlds and human-robot-teams: empirical studies towards a new educational age," *2016 IEEE Global Engineering Education Conference (EDUCON)*, Abu Dhabi, United Arab Emirates, pp. 142–149, 2016.

Samoilă, C., Ursuţiu, D., Jinga, V., "The remote experiment compatibility with Internet of Things," *13th International Conference on Remote Engineering and Virtual Instrumentation (REV)*, Madrid, Spain, pp. 204–207, 2016.

Sandu, F., Costache, C., Balan, T., "Semantic data aggregation in heterogeneous learning environments," *IEEE 21st International Symposium for Design and Technology in Electronic Packaging (SIITME)*, Brasov, Romania, pp. 409–412, 2015.

Sarıta¸s, M. T., "The emergent technological and theoretical paradigms in education: the interrelations of cloud computing (CC), connectivism and Internet of Things (IoT)," *Acta Polytechnica Hungarica*, vol. 12, no. 6, 2015, pp. 161–179.

Shi, Y., Qin, W., Suo, Y., Xiao, X., "Smart classroom: bringing pervasive computing into distance learning." In: Nakashima H., Aghajan H., Augusto J.C. (eds) *Handbook of Ambient Intelligence and Smart Environments*, pp. 881–910. Springer, Boston, MA, 2010.

Shirehjini, A. A. N., Yassine, A., Shirmohammadi, S., Rasooli, R., Arbabi, M. S., "Cloud assisted IoT based social door to boost student-professor interaction," *International Conference on Human-Computer Interaction*, pp. 426–432, Toronto, 2016.

Srivastava, A., Yammiyavar, P., "Augmenting tutoring of students using tangible smart learning objects: an IoT based approach to assist student learning in laboratories," *International Conference on Internet of Things and Applications (IoTA)*, Pune, India, pp. 424–426, 2016.

Sula, A., Spaho, E., Matsuo, K., Barolli, L., Miho, R., Xhafa, F., "An IoT-based system for supporting children with autism spectrum disorder," *2013 Eighth International Conference on Broadband and Wireless Computing, Communication and Applications*, Compiegne, France, pp. 282–289, 2013.

Sula, A., Spaho, E., Matsuo, K., Barolli, L., Miho, R., Xhafa, F., "A smart environment and heuristic diagnostic teaching principle-based system for supporting children with autism during learning," *28th International Conference on Advanced Information Networking and Applications Workshops*, Victoria, BC, Canada, pp. 31–36, 2014.

Tan, W., Chen, S., Li, J., Li, L., Wang, T., Hu, X., "A trust evaluation model for E-learning systems," *Systems Research and Behavioral Science*, vol. 31, no. 3, 2014, pp. 353–365.

Tech, E., "Internet of Things and the humanization of healthcare technology," 2015. https://vz.to/2LWy3KJ.

Tunc, C., Hariri, S., Montero, F. D. L. P., Fargo, F., Satam, P., "CLaaS: Cybersecurity Lab as a service–design, analysis, and evaluation," *International Conference on Cloud and Autonomic Computing*, Boston, MA, pp. 224–227, 2015.

Ueda, T., Ikeda, Y., "Stimulation methods for students' studies using wearables technology," *IEEE Region 10 Conference (TENCON)*, Singapore, pp. 1043–1047, 2016.

UNESCO, "Teaching and learning: achieving quality for all," 2014. https://bit.ly/2X1ZaF6.

Uskov, V., Pandey, A., Bakken, J. P., Margapuri, V. S., "Smart engineering education: the ontology of Internet-of-Things applications," *IEEE Global Engineering Education Conference (EDUCON)*, Abu Dhabi, United Arab Emirates, pp. 476–481, 2016.

Uzelac, A., Gligoric, N., Krco, S., "A comprehensive study of parameters in physical environment that impact students' focus during lecture using Internet of Things," *Computers in Human Behavior*, vol. 53, 2015, pp. 427–434.

Valpreda, F., Zonda, I., "Grüt: a gardening sensor kit for children," *Sensors*, vol. 16, no. 2, 2016, p. 231.

Voas, J., "Networks of 'things'," *NIST Special Publication*, vol. 800, no. 183, 2016, pp. 800–183.

Wan, R., "Network interactive platform ideological and political education based on internet technology," *International Conference on Economy, Management and Education Technology*, Chongqing, 2016.

Wang, J., "The design of teaching management system in universities based on biometrics identification and the Internet of Things technology," *International Conference on Computer Science & Education (ICCSE)*, Cambridge, pp. 979–982, 2015.

Wang, Y., "English interactive teaching model which based upon Internet of Things," *International Conference on Computer Application and System Modeling (ICCASM 2010)*, Taiyuan, China, vol. 13, pp. 513–587, 2010.

Wang, Y., "The construction of the psychological health education platform based on Internet of Things," *Applied Mechanics and Materials*, vol. 556, 2014, pp. 6711–6715.

Weber, R. H., "Internet of Things–new security and privacy challenges," *Computer Law & Security Review*, vol. 26, no. 1, 2010, pp. 23–30.

Xue, Y.-F., Sun, H.-L., Wu, Y.-H., Chen, S.-K, "Design and development of digital teaching management system based on Internet of Things," *International Symposium on Computer Science and Society*, Kota Kinabalu, Malaysia, pp. 138–141, 2011.

Yin, C., Dong, Y., Tabata, Y., Ogata, H., "Recommendation of helpers based on personal connections in mobile learning," *IEEE Seventh International Conference on Wireless, Mobile and Ubiquitous Technology in Education*, Takamatsu, Japan, pp. 137–141, 2012.

Zhang, N., "A campus big-data platform architecture for data mining and business intelligence in education institutes," *6th International Conference on Machinery, Materials, Environment, Biotechnology and Computer*, Tianjin, 2016.

Zhu, L., "Research and design of the future classroom based on big data and cloud processing," *International Conference on Audio, Language and Image Processing (ICALIP)*, Shanghai, China, pp. 111–114, 2016.

8 IoT Education[1]

8.1 INTRODUCTION

By 2025 there will be 75 billion connected smart devices (Newman, 2020). Every industry is realizing Internet of Things (IoT) can help make their systems and devices more effective and efficient, including healthcare, education, retail, logistics, workforce management, and more. Thus, there is a clear urgency for an IoT workforce that has the skills and knowledge in the areas of computer science, engineering, and cybersecurity to build the IoT devices and systems. A quick search on *indeed.com* for US jobs in this area produced 25,920 jobs when searching for 'Internet of things'. Other search terms produced 16,840 jobs with 'IoT'; 5,314 with 'IoT Systems Engineer'; and 7,125 jobs with 'IoT Engineering'. It is important to note that those results did not include open jobs in the areas of data analytics, algorithms, machine learning, or security, which are clearly important disciplines in the design and implementation of IoT as well. In addition, the Bureau of Labor and Statistics predicts a 30% increase in these jobs by 2026.

In this book, we cover the major areas that are using IoT (e.g., education, cities, homes, healthcare) – but we need to address how we will teach the next generation of engineers that analyze, design, and implement these complex systems. We need a new engineering discipline for the 21st century such as IoT engineering, IoT systems engineering, or network engineering. This is similar to past innovations that called for new engineering disciplines such as when electrical engineering emerged in the late 19th century with the invention of the electric motor as well as when the Industrial Revolution called for a chemical engineering discipline to design the mass production of chemicals. Even more recently, we have experienced the Software Engineering discipline that emerged with the need for the development of complex software systems. The industry needs a complete cyber-physical system (CPS)/IoT curriculum that will require core set of competencies from multiple disciplines that is already existing in our engineering and computer science programs.

The goal of this chapter is to present an analysis that determines the state of this educational need and to provide the knowledge to advance the effort in adapting IoT training within academic institutions to meet the urgent IoT employment needs. This chapter presents an analysis of courses related to educating students to build IoT and CPS and the results of a program review for CPS/IoT-related courses currently being offered at the top 50 universities ranked by Collegechoice.net. The results from this review were analyzed and mapped to the NIST networks of things' (NoT) primitives discussed earlier in this book and the Association for Computing

---

[1] Excerpts of this chapter are from DeFranco, Kassab and Voas (2018); DeFranco & Kassab "Considerations for an Internet of Things Curriculum," HICSS 2019.

DOI: 10.1201/9781003027799-8

Machinery/Institute of Electrical and Electronic Engineers (ACM/IEEE) computer science knowledge areas (KAs). In addition, to highlight specific course projects, this chapter will be able to assist in the effort of building or adapting current academic programs offered to meet the current and future needs IoT/CPS development.

8.2 INTERNET OF THINGS

Since we are mapping the courses to the NIST NoT primitives discussed earlier in the book, a review of those primitives is useful here. Recall that an IoT is representation of a NoT, and because of the upward trend of IoT applications and systems in various domains, it is useful to have a consistent language to build NoTs (Voas, 2016). As described in Chapter 2, IoTs can be described by five primitives (i.e., the five Lego-like building blocks for any IoT-based system) as proposed in the NIST Special Publication 800-183 (Voas, 2016):

1. A sensor is an electronic utility (e.g., cameras and microphones) that measures physical properties such as sound, weight, humidity, temperature, and acceleration. Properties of a sensor could be the transmission of data (e.g., RFID), Internet access, and/or be able to output data based on specific events.
2. A communication channel is the medium by which data is transmitted (e.g., physical via Universal Serial Bus (USB), wireless, wired, verbal).
3. An aggregator is software that is based on mathematical function(s). This software can transform groups of raw data (from any source) into intermediate, aggregated data. Aggregators have two actors for consolidating large volumes of data into lesser amounts:
   a. Cluster is "an abstract grouping of sensors (along with the data they output) that can appear and disappear instantaneously".
   b. Weight is "the degree to which a particular sensors data will impact an aggregator's computation".
4. A decision trigger "creates the final result(s) needed to satisfy the purpose, specification and requirements of a specific NoT". A decision trigger is a conditional expression that triggers an action and abstractly defines the end purpose of a NoT. A decision trigger output can control actuators and transactions.
5. External utility (eUtility) is a "hardware product, software or service which executes processes or feeds data into the overall data flow of the NoT".

Note, that any specifically purposed NoT may not include all five primitives. For example, you could have a NoT without any sensors. The simplest way to think about any IoT is that the "things" are what make IoT unique. Remember, in the IoT is an acronym, the (T) is the letter that matters the most. The primitives are essentially the (T)s. We need to educate the next generation of computer scientists and engineers on the (T)s (DeFranco, Kassab & Voas, 2018).

8.3 THE ACM AND IEEE COMPUTER SCIENCE KNOWLEDGE AREAS

The ACM and IEEE Computer Society have a long record of sponsoring efforts to establish international curricular guidelines for academic programs in computing. These guidelines are on approximately a 10-year cycle starting with the publication of *Curriculum 68* (Atchison et al., 1968) 50 years ago. The latest IEEE/ACM Curriculum Guidelines for Undergraduate Degree Programs in Computer Science was published in 2013 (Force, 2013). In Computer Science, one can view the body of knowledge (BOK) as a specification of the content to be taught and a curriculum as an implementation. As shown in Table 8.1, the IEEE/ACM BOK is composed of 18 KAs.

KAs are not intended to be in one-to-one correspondence with a course that is part of a curriculum. A curriculum should have courses that integrate topics from multiple KAs. A set of topics were recommended from Voas and Laplante to be considered when creating new computer science curricula, or the recommendations can be used to modify an existing computer science curriculum (Voas & Laplante, 2017). Further, if at the curriculum is focused on CPS issues rather than IoT, modifying a systems engineering, electrical engineering, or mechanical engineering curricula might be worth pursing as well.

8.4 MAPPING STUDY AND PROGRAMS REVIEW

We conducted a review with the goal of discovering how IoT knowledge is currently being taught in academia. This review was completed in two parts:

1. *A Mapping Study*: First we searched the Engineering Village database to learn about existing research and the proposed courses for designing and implementing IoT and CPS. The Engineering Village database is a comprehensive database that aggregates 12 engineering literature and patent databases to provide coverage from a wide range of trusted engineering sources.

TABLE 8.1

IEEE/ACM Computer Science Knowledge Areas

1	Algorithms and complexity	10	Networking and communications
2	Architecture and organization	11	Operating systems
3	Computational science	12	Platform-based development
4	Discrete structures	13	Parallel and distributed computing
5	Graphics and visualization	14	Programming languages
6	Human–computer interaction	15	Software development fundamentals
7	Information assurance and security	16	Software engineering
8	Information management	17	Systems fundamentals
9	Intelligent systems	18	Social issues and professional practice

2. *Programs Review*: In the second part of this review, we analyzed CPS and IoT-related programs at the top 50 universities as ranked by Collegechoice.net and TopUniversities.com. Collegechoice.net is an aggregate of US News & World Report and the National Center for Education Statistics, and TopUniversities. com includes international universities for IoT and CPS course offerings.

At this point, it is important to differentiate in order to not confuse "IoT in Education" and "IoT Education". IoT in Education was covered in another chapter of this book and focuses on technological tools that improve academic infrastructure. This discussion, IoT Education, focuses on courses or subjects from which to teach fundamental concepts of computer science (Elyamany & AlKhairi, 2015). An example of the IoT Education is a focus on utilizing IoT in an education setting such as a tool to manage student attendance with RFID readers at the school, library, cafeteria, dormitory entrances, and to monitor student activities in these locations (Caţă, 2015). In this chapter, we are focusing on IoT Education. Thus, during the first part of the study, the research that focused on the first facet (IoT *in* education) were excluded. That said, there are much fewer research papers presenting studies that matched our criteria in this mapping study which focuses on educating the next generation of engineers and computer scientists to design and build IoT and CPS.

We searched the database using keywords related to IoT such as "Internet of Things", "IoT", "Network of Things" and "NOT", "Cyber Physical Systems". We also included a string made up of keywords-related education such as "education" and "course". An example of a search performed was ("Internet of Things" OR "IoT" OR "Network of Things" OR "NoT" OR "Cyber Physical Systems") AND ("Education" OR "Course").

This search resulted in 309 articles. To determine whether a study should be included, the following exclusion criteria were used:

1. A study that was not a peer reviewed such as an opinion, viewpoint, keynote, discussion, editorial, tutorials, prefaces, and presentations;
2. Studies that were not in English;
3. Studies that focused on IoT in Education but not IoT Education as we discussed previously.

After reviewing the titles and metadata of the research, 54 papers were downloaded to be considered in this analysis. After reading the abstract of each paper, there were 30 articles rejected on the basis of the exclusion criteria mentioned earlier. The remaining 24 papers were read in their entirety, and ultimately, only 11 studies remained to be considered that meet the inclusion criteria.

While reading the papers and reviewing the courses retrieved from the two parts of the study, the primitives and Computer Science KAs were documented. The findings are discussed in the next few sections.

8.4.1 Overview of Courses from the Mapping Study

Nearly half of the courses discussed in the 11 studies identified are US based with others located in other countries (Italy, Spain, India, UK, and China). Eight

TABLE 8.2
List of Courses Extracted from the Mapping Study

	Course Name	Article Source
1	Ambient Intelligence System Design	Corno and DeRussis (2017)
2	Systems Administration	Gonzalez-Nalda et al. (2014)
3	Cyber-Physical Systems	Möller and Vakilzadian (2016)
4	Internet of Things	Zhong and Liang (2016)
5	Ambient Intelligence: Technology and Design	Corno, De Russis, and Bonino (2016)
6	Smart Sensors and Internet of Things	Islam, Mukhopadhyay, and Suryadevara (2017)
7	My Digital Life	Kortuem et al. (2013)
8	A CPS Project in a Microprocessor System Design	Crenshaw (2013)
9	Pervasive Computing Systems	Graves et al. (2015)
10	Cyber-Physical Systems	Zalewski and Gonzalez (2017)
11	Embedded Systems Design	Hamblen and Van Bekkum (2013)

of the courses (all shown in Table 8.2) were delivered as undergraduate students. Interestingly, eight of these courses were part of *electrical engineering* and *electrical and computer engineering* programs. The remaining four courses were part of *engineering science, computer and information science*, and *software engineering* programs.

8.4.2 OVERVIEW OF COURSES FROM THE PROGRAMS REVIEW

CPS/IoT-related course offerings at the top 50 US-based and international universities ranked by collegechoice.net and TopUniversities.com were also reviewed. Relevant programs for courses with content focused on the understanding, design, and/or implementation of IoT, CPS, and NoT were searched. Table 8.3 shows the breakdown of the results. Twenty-eight of those universities had graduate courses with a CPS/IoT focus, eight courses were offered as both undergraduate and graduate courses, and more than half of those courses were taught in electrical and computer engineering programs (DeFranco, Kassab & Voas, 2018). A complete list of courses/universities we reviewed are available via the link: goo.gl/UUtN8T.

TABLE 8.3
Number of IoT/CPS Courses at the Top 50 Ranked Universities (DeFranco, Kassab & Voas, 2018)

Universities with IoT/CPS Courses	Total IoT/CPS Courses	Undergraduate Courses	Graduate Courses
28	45	18	35

8.5   CPS/IoT COURSES AND THE IoT PRIMITIVES

Next, course descriptions and course structures were reviewed to determine the IoT primitive focus. Again, these courses were taught at the top 50 universities. We analyzed the course syllabi, descriptions, and available materials for each of the 45 courses. The resulting course data was mapped to the NoT primitives. For example, in an "embedded systems" course, the description stated:

> Lectures will cover theoretical concepts of embedded and **cyberphysical systems** including discrete and continuous dynamics, hybrid systems, state machines, concurrent computation, embedded systems architecture and scheduling. Lab involves programming embedded applications for the **decentralized software services** architecture using C# and the Microsoft **Robotics** Software Development Kit (SDK) together with the hardware image processing and tracking capabilities of the Kinect **sensor**.

This description was mapped accordingly to the "sensor", "communication", and "decision trigger" IoT primitives. The complete mapping matrix for the reviewed 45 courses to the IoT primitives is available via the link: goo.gl/8VmGTV. The results showed that only 11% of the 45 courses seem to cover all five primitives (see Figure 8.1).

The courses, Interconnected Embedded Systems, Networked CPSs, IoT Intelligent and Connected Systems, and Body Sensor Networks in the IoTs appear to be introductory to IoT and CPS technical and design understanding.

When the combined list of courses from top 50 universities and the 11 courses from the retrieved studies were analyzed, we determined that the primitive focus that was most prevalent was "eUtility" with almost 25% of the courses covering that primitive in its content. In general, the courses described had various scenarios to teach each NoT primitive. Particularly, courses that covered the NoT primitives involved projects that focus on the design and implementation of devices for

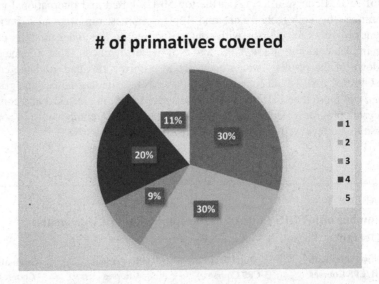

FIGURE 8.1   Number of Internet of Things (IoT) primitives covered in the courses at the top 50 ranked universities.

remote-controlled systems or smart home devices. For example, a course on ambient intelligence (AMI) would provide students access to smart home devices where they could learn to design and implement an AMI system. An AMI system could involve sensing, reasoning about the sensed data (aggregation) on the cloud (eUtility), communicating the data using web protocols and architectures, and actuation (decision trigger) and interaction with the user (Corno & De Russis, 2017). Another course may focus on the design of a generic remote-controlled system which may utilize robot cameras and gyroscopes (sensors) on mobile phone (eUtility), wireless communication of the data to and from robot, aggregating and reasoning about the data to perform robot movement (actuation) (Gonzalez-Nalda et al., 2014). A CPS project example focused on automation of industrial processes using an autonomous-guided vehicle (AGV). The requirements for an AGV system include sensors to collect data that could inform if the AGV is near a start and stop point by communicating and aggregating data stored on a cloud or server (eUtility) and triggering a decision (Möller & Vakilzadian, 2016).

8.6 CPS /IoT COURSES AND KNOWLEDGE AREAS OF COMPUTER SCIENCE

Most of the analyzed courses focused on IoT and CPS development. For example, an AMI system utilizes pervasive technology and refers to the capacity of an IoT system to sense the environment and to respond to the presence of people. Pervasive computing are the microprocessors that facilitate the consistent communication connection in AMI systems.

In other words, AMI builds upon pervasive computing and human-centric computer interaction design (KA: human–computer interaction). Environments that use AMI have the following properties (Lee, 2017):

- Identity recognition;
- Awareness of an individual's presence;
- Context awareness (e.g., traffic, weather);
- Activity recognition and;
- Individual needs adaptation.

An excellent example of AMI are Ambient Assistant Living (AAL) technologies. These are technologies designed to assist autonomous aging such as detecting falls or medication reminders (Jaschinski & Allouch, 2015). Another example of an ambient technology device is one that detects and absorbs emission carbon dioxide in order to reduce control global warming (Manoj & Renold, 2011).

Table 8.4 shows the analysis of course and project descriptions. The courses were mapped to the IEEE/ACM computer science KAs. Noting that although there does not seem to be a pattern or consistency in the KAs covered, it is clear that the IoT/CPS skill set matches well with the IEEE/ACM computer science KAs.

The KAs for the courses were determined by the learning objectives and/or projects depending on what was provided by the research. For example, a learning objective for the Microprocessor Systems Design course was to write a pair of networked

TABLE 8.4

Course Mapping to IEEE/ACM Computer Science Knowledge Areas

Course Name Article Source	Computer Science Knowledge Areas
Ambient Intelligence System Design (Corno & De Russis, 2017)	(2) (6) (9) (10) (12) (13) (14) (15) (16)
Systems Administration (Gonzalez-Nalda et. al., 2014)	(1) (2) (3) (5) (9) (10) (11) (12) (14) (15) (17)
Cyber-Physical Systems (Möller & Vakilzadian, 2016)	(1) (2) (3) (7) (8) (9) (10) (12) (13) (16) (18)
Internet of Things (Zhong & Liang, 2016)	(10) (12) (14) (15) (17)
Ambient Intelligence: Technology and Design (Corno, De Russis & Bonino, 2016)	(6) (8) (10) (12) (14) (15) (16) (18)
Smart Sensors and Internet of Things (Islam, Mukhopadhyay & Suryadevara, 2017)	(2) (7) (10) (12) (15)
My Digital Life (Kortuem et al., 2013)	(1) (2) (15) (18)
A CPS Project in a Microprocessor System Design (Crenshaw, 2013)	(2) (9) (10) (12) (15)
Pervasive Computing Systems (Graves et. al., 2015)	(8) (9) (10) (11) (12) (15) (16)
Cyber-Physical Systems (Zalewski & Gonzalez, 2017)	(15) (16) (18)
Embedded Systems Design (Hamblen & Bekkum, 2013)	(9) (10) (12) (15) (16)

software applications that exchange data between the embedded system and a computer. This objective clearly maps to KAs: *networking and communications* and *software development fundamentals*.

In general, the course objectives were to have the students learn to design and manage complex/distributed systems (*KA: parallel and distributed computing*). Many of the projects assigned in the courses focused on practical programming (*KA: software development fundamentals and algorithms and complexity*) of the IoT/CPS. In another project, the students developed an Android-based remote system for teleoperating a mobile robot controlled by means of a Raspberry Pi. The software developed by the students used the mobile phone gyroscope which translated the left and right tilt movements of the smartphone into motion commands to be followed by a robot. Other interesting projects were in the area of smart things used in cities and homes. There was also an environmental monitoring (e.g., air, water, and soil quality; health quality; monitoring flood; volcanic eruptions; and traffic) project.

8.7   RECOMMENDATIONS

It is urgent that established engineering programs provide engineering students with the skills, tools, and training to design, implement, verify, and validate these complex IoT and CPSs. The current state according to the data presented in this chapter indicates that course offerings in this area are at the beginning stages. This is not surprising considering the challenge academic programs have in adding courses and/or establishing new programs in general. Thus, it may be more practical for a traditional college-level curriculum to adapt the conventional courses to include the tools and training that address the need to develop this skill set of next-generation engineers in designing and building IoT and CPS. For example, an effective approach is to create learning modules that would address specific learning objectives. The modules would highlight CPS/IoT concepts using appropriate tools and hands-on exercises that could easily fit into existing courses such as embedded systems, systems administration, computer security, software architecture, and software construction. Virginia Tech created CPS security-focused learning modules for this purpose (Deshmukh, Patterson & Baumann, 2016). Another approach is to create elective courses where students implement practical CPS or IoT systems as a course project similar to those described earlier.

Accordingly, the path of least resistance to create new programs is by modifying existing programs since many of these courses may already exist at the institution and only require slight modifications. This appears to be the most efficient way to create new CPS/IoT educational program that is relevant, timely, and available now.

The burden and challenge is on higher education to prepare the next generation of engineers with the skill set necessary to build these complex systems. Universities are being tasked to provide working knowledge of a new engineering discipline that adapts the IoT and CPS differences in existing engineering disciplines. This is not an easy mission as it is challenging for educators to keep up with the rapid technology development pace as well as impart this knowledge onto students (Zhu et al., 2014).

REFERENCES

Atchison, W. F., Conte, S. D., Hamblen, J. W., Hull, T. E., Keenan, T. A., Kehl, W. B., McCluskey, E. J., Navarro, S. O., Rheinboldt, W. C., Schweppe, E. J., Viavant, W., "Curriculum 68: recommendations for academic programs in computer science: a report of the ACM curriculum committee on computer science," *Communications of the ACM*, vol. 11, no. 3, 1968, pp. 151–197.

Caţă, M., "Smart University, a new concept in the internet of things," *2015 14th RoEduNet International Conference-Networking in Education and Research (RoEduNet NER)*, Craiova, Romania, pp. 195–197, 2015.

Corno, F., De Russis, L., "Training engineers for the ambient intelligence challenge," *IEEE Transactions on Education*, vol. 60, no. 1, 2017, pp. 40–49.

Corno, F., De Russis, L., Bonino, D., "Educating internet of things professionals: the ambient intelligence course," *IT Professional*, vol. 18, no. 6, 2016, pp. 50–57.

Crenshaw, T. L., "Using robots and contract learning to teach cyber-physical systems to undergraduates," *IEEE Transactions on Education*, vol. 56, no. 1, 2013, pp. 16–120.

DeFranco, J., Kassab, M., Voas, J., "How do you create an internet of things workforce?," *IT Professional*, vol. 4, 2018, pp. 8–12.

Deshmukh, P. P., Patterson, C. D., Baumann, W. T., "A hands-on modular laboratory environment to foster learning in control system security," *2016 IEEE on Frontiers in Education Conference (FIE)*, Erie, PA, pp. 1–9, 2016.

Elyamany, H. F., AlKhairi, A. H., "IoT-academia architecture: a profound approach," *2015 16th IEEE/ACIS International Conference on Software Engineering, Artificial Intelligence, Networking and Parallel/Distributed Computing (SNPD)*, Takamatsu, Japan, pp. 1–5, 2015.

Force, A. J. T., "Computer science curricula 2013: curriculum guidelines for undergraduate degree programs in computer science," Technical Report, Association for Computing Machinery (ACM) IEEE Computer Society, 2013.

Gonzalez-Nalda, P., Calvo, I., Etxeberria-Agiriano, I., García-Ruíz, A., Martíinez-Lesta, S., Caballero-Martín, D., "Building a CPS as an educational challenge," *International Journal of Online Engineering (iJOE)*, vol. 10, no. 4, 2014, pp. 52–58.

Graves, C. A., Negron, T. P., Chestnut II, M., Popoola, G., "Studying smart spaces using an 'embiquitous' computing analogy," *IEEE Pervasive Computing*, vol. 14, no. 2, 2015, pp. 64–68.

Hamblen, J. O., Van Bekkum, G. M., "An embedded systems laboratory to support rapid prototyping of robotics and the Internet of Things," *IEEE Transactions on Education*, vol. 56, no. 1, 2013, pp. 121–128.

IEEE, "IEEE Internet of Things," 2021. http://iot.ieee.org/about.html.

Islam, T., Mukhopadhyay, S. C., Suryadevara, N. K., "Smart sensors and Internet of Things: a postgraduate paper," *IEEE Sensors Journal*, vol. 17, no. 3, 2017, pp. 577–584.

Jaschinski, C., Allouch, S. B., "An extended view on benefits and barriers of ambient assisted living solutions," *International Journal of Advanced Life Sciences*, vol. 7, no. 2, 2015, pp. 40–53.

Kortuem, G., Bandara, A. K., Smith, N., Richards, M., Petre, M., "Educating the Internet-of-Things generation," *Computer*, vol. 46, no. 2, 2013, pp. 53–61.

Lee, F., "Ambient intelligence the ultimate IoT use cases," 2017. https://www.iotforall.com/ambient-intelligence-ami-iot-use-cases/.

Manoj, A. P., Renold, A. P., "Pervasive ambient intelligence system: a zigbee based sensor networks for ambient monitoring," *2011 International Conference on Signal Processing, Communication, Computing and Networking Technologies (ICSCCN)*, Thuckalay, India, pp. 619–622, 2011.

Möller, D. P., Vakilzadian, H., "Technology-enhanced learning in cyber-physical systems embedding modeling and simulation," *International Journal of Quality Assurance in Engineering and Technology Education (IJQAETE)*, vol. 5, no. 3, 2016, pp. 32–45.

Newman, D., "5 IoT trends to watch in 2021," Forbes, November 25, 2020.

Voas, J., "Networks of things," *NIST Special Publication*, vol. 800, 2016, p. 183.

Voas, J., Laplante, P. A., "Curriculum considerations for the Internet of Things," *Computer*, vol. 50, no. 1, 2017, pp. 72–75.

Zalewski, J., Gonzalez, F., "Evolution in the education of software engineers: online course on cyberphysical systems with remote access to robotic devices," *International Journal of Online Engineering (iJOE)*, vol. 13, no. 8, 2017, pp. 133–146.

Zhong, X., Liang, Y., "Raspberry Pi: an effective vehicle in teaching the internet of things in computer science and engineering," *Electronics*, vol. 5, no. 3, 2016, p. 56.

Zhu, R., Lei, J., Mao, T., Zhou, B., Guo, M., "Innovative network engineering practice based on multimedia education scheme," *Journal of Multimedia*, vol. 9, no. 3, 2014, pp. 463–468.

9 IoT in Healthcare[1]

9.1 INTRODUCTION

Healthcare is a data-intensive discipline (Patil & Seshadri, 2014) in which large-scale data is generated, disseminated, stored, and accessed daily. In 2017, 16.5 million patients globally exploited remote health monitoring (a 41% growth from 2016), and this panorama has the potential to reach 50.2 million by 2021 (Mack, 2017). Also, since January 1, 2018, the US Centers for Medicare & Medicaid Services developed new reimbursement incentives to promote the adoption of "active feedback loop" devices to provide real-time monitoring (Daniel & Uppaluru, 2017). Data created when a patient is monitored or undergoes some tests need to be stored in order to be accessible at a later time by a healthcare provider within the same or even a different network or context. Healthcare applications that are connected to the Internet, also referred to as Internet of Things (IoT) applications in healthcare, have been widely forecast, investigated, and even deployed on a small scale. For example, some hospitals have begun implementing "smart beds" that can detect when they are occupied and when a patient is attempting to get up, sending this information over the network/Internet to nurses (Babu & Jayashree, 2015). The beds can also self-adjust to ensure that appropriate pressure and support is applied to the patient without having to be manually adjusted by the nurses. Another area where smart technology is being discussed as an asset is coupled with home medication dispensers to automatically upload data to the cloud when medication is not taken or any other indicators for which the care team should be alerted (Chouffani, 2016).

The nature of healthcare and the computational and physical technologies and constraints present several challenges to system designers and implementers. These challenges are complex and include the following concerns:

1. Physical (e.g., available technology);
2. Logical (e.g., analytics, languages, tools);
3. Political (e.g., funding, mandates);
4. Behavioral (i.e., desired functionality);
5. Communications (e.g., available channels);
6. Ethical (e.g., governmental privacy protection standards);
7. Structural (e.g., patterns of architecture and design).

Despite these concerns, polls show that 90% of Americans still value online access to their health records (FDA, 2013). It is easy to perceive that IoT can contribute to enhancing the quality of caregiving for patients at a reduced cost. For example, it was projected that by 2020, the number of Americans who are expected to need

---

[1] Excerpts of this chapter are from Laplante, P. A,, Kassab, M., Laplante, N., Voas, J. "Building Caring Healthcare Systems in the Internet of Things," *IEEE Systems Journal*, 2017.

DOI: 10.1201/9781003027799-9

assistance of some kind to be 117 million, yet the overall number of unpaid caregivers (e.g., family members) is only expected to reach 45 million. That makes one unpaid caregiver for every 2.6 persons needing assistance.

Therefore, a large market opportunity is presented by those people who are online and connected, and who would make use of technology that is intuitive and consumer-friendly to provide care. Yet, there is not enough technology that can meet caregiving needs.

According to a recent study conducted by Project Catalyst and the Health Innovation Technology Laboratory (HITLAB) to better understand how caregivers are currently using technology, an average of 71.5% of caregivers reported that they are interested in using technology across 17 tested care-giving tasks if such technology exists (Caregivers & Technology, 2016).

In this chapter, we present a discussion on using IoTs for healthcare. We will also present a discussion on how IoT has been used for digital surveillance to combat the coronavirus disease 2019 (COVID-19) pandemic. We also discuss considerations for traditional and emerging quality requirements in IoTs for healthcare.

9.2 GENERAL CLASSIFICATION FOR USE CASES FOR IoT IN HEALTHCARE

Healthcare can be delivered in three broad-based setting types:

1. Acute care,
2. Community-based care, and
3. Long-term care.

Acute care refers to a hospital setting where the caregivers are paid healthcare professionals. Community-based care is delivered in a home setting, where the patient is living in his or her own or another's home and where caregivers are either paid professionals or unpaid family members or friends. Long-term care refers to nursing homes or other skilled nursing facilities where patients reside for weeks, months, years, or for the remainder of their lives and where caregivers are paid professionals.

IoTs can be used to collect patient and other data in these settings and aggregate the data using analytics and then reporting this information to caregivers and/or take some action (such as shutting down a faulty medical device). It would be futile to try to enumerate all conceivable IoT applications in healthcare since after completing any list, new applications will be innovated.

Instead, we define three classes of use cases of healthcare IoTs:

Class A: tracking humans (e.g., patients, caregivers, and family members),
Class B: tracking things (e.g., medical devices, supplies, and specimens), and
Class C: tracking humans and things.

Taking the dimensions of care settings and IoT application classes yields nine general use cases: acute (A, B, C), long term (A, B, C), and home (A, B, C). In the next section, we will focus on the first class: tracking humans.

9.3 IoT FOR TRACKING HUMANS

Tracking humans involves tracking humans' data (e.g., patients, caregivers, family members) using IoT devices. Perhaps the most mature field for IoT in healthcare is patient data gathering. Currently, telemetry monitors can automatically measure and send or upload electrocardiogram (EKG) stats, core body temperature, blood pressure, urine output, etc.

By monitoring these vital signs, healthcare professionals can detect and start care earlier for infectious disease, cancer, heart failure, etc. Another example in this class involves tracking the physical location of patients in any setting (acute, long term, home). From tracking wandering patients admitted to emergency room (ER) to tracking patients with dementia, the IoT could geolocate patients with Alzheimer's disease or self-destructive behaviors such as bulimia, cutting, or suicidal tendencies. Such tracking can already be accomplished with commercial GPS bracelets, but local proximity sensors connected through the Internet or cloud-based technologies could allow tracking inside of the facility or home or outside these, where GPS signals may not reach.

To demonstrate tracking humans, we will present in the next subsection a case study on tracking patients with alcohol withdrawal syndrome (AWS) symptoms.

9.3.1 ALCOHOLISM USE CASE

Alcoholism is a long-term chronic disease in which a person has developed an unhealthy dependence on alcohol. In the United States, there are close to 14 million people who are either alcohol abusers or alcoholics (National Institute of Alcohol Abuse and Alcoholism, 2016). Fortunately, no matter how severe the problem may seem, most people with an alcohol use disorder can benefit from some form of treatment. Research shows that about one-third of people who are treated for alcohol problems have no further symptoms 1 year later. Many others substantially reduce their drinking and report fewer alcohol-related problems.

In the early stages of the treatment phase, a patient may suffer from AWS, which refers to the set of symptoms that occur when a heavy drinker suddenly stops or significantly reduces their alcohol intake. With AWS, a patient may experience a combination of physical and emotional symptoms which include one or more of the following:

Anxiety or jumpiness
Depression
Shakiness or trembling
Irritability
Sweating
Fatigue
Nausea and vomiting
Loss of appetite
Insomnia
Headache

Some symptoms of AWS can be as severe as hallucinations and seizures. At its most extreme, AWS can be life-threatening. Detecting the degree of severity of these symptoms is essential to adjust the treatment. Matching the right therapy to the individual is important to its success. No single treatment will benefit everyone in the alcoholism case.

Many of the above AWS symptoms could potentially be monitored using a specialized IoT or non-Internet-enabled analytics. For example, a patient with AWS needs to be carefully monitored regarding trembling and irregular movement. Sensors can be strategically placed in the patient's home and used to pick up on accelerated and irregular walking or movement activity as compared to walking or moving at a normal pace.

In addition, a patient can be monitored for episodes of vomiting by observing instances of bathroom use via an IoT. A sensor that can detect the odor of vomit could provide additional cues in the diagnosis and management of the AWS patient in home care settings.

An IoT system can render a decision on the existence of AWS symptoms and the degree of such symptoms. If a patient has mild-to-moderate withdrawal symptoms, a healthcare provider may prefer to continue the treatment in an outpatient setting while prescribing some medications to reduce the severity of the symptoms, especially if a patient has supportive family and friends. If the symptoms are extremely severe, then the system may alert the case as a medical emergency that requires an acute setting. Table 9.1 depicts a simple construct for an AWS patient in a long-term or home care setting using an IoT.

9.3.2 Digital Surveillance to Combat COVID-19

The novel coronavirus (SARS-COV-2) that causes the COVID-19 has been spreading worldwide at an accelerated rate. The situation was declared a pandemic by the World Health Organization (WHO) on March 11, 2020. Almost immediately, governments around the world started to pursue containment measures to help slow the spread of the virus.

TABLE 9.1

Alcohol Withdrawal Syndrome (AWS) Use Case Construct

Model Element	Realization
Sensor	Proximity sensor(s)
Snapshot (time)	Once per minute
Cluster	Set of (three) proximity sensors per room or hallway
Aggregator	Determine severity of AWS symptom
Weight room	Layout dependent
Communication channel	ZigBee1 compliant network of sensors or clusters or aggregator, wired (Internet) to eUtility
eUtility	Remote monitoring software (onsite – e.g. administration desk)
Decision	Degree of existing AWS symptoms

Digital surveillance has been deployed as part of these measures to:

   i. Track confirmed and potentially impacted cases with the virus,
   ii. Enforce lockdown when necessary, and
   iii. Generate a much-needed source of data and statistics to the authorities.

The IoT community has been stimulated with the COVID-19 outbreak to conceive solutions to combat the pandemic. IoT protocols (e.g., Bluetooth Low Energy (BLE), Near-Field Communications (NFC), Radio Frequency Identification (RFID), global positioning system (GPS), and Wi-Fi) are receiving great attention for providing solutions spanning the spectrum from a biosensor capable of detecting the SARS-COV-2 virus in the air (Qiu et al., 2020) to the rapidly emerging digital surveillance technologies to track individuals and crowds. A digital surveillance for an individual relies on the unique identifiers that are either temporally (RFID tags) or permanently assigned to a person like personal identifiers, and a way to sample individual's locations, either on short or long temporal scales for authorities to keep track of the citizens or temporary residents.

At the most general level, surveillance of humans can be defined as "regard or attendance to others (whether a person, a group, or an aggregate as with a national census) or to factors presumed to be associated with these. A central feature is gathering some form of data connectable to individuals (whether as uniquely identified or as a member of a category)" (Marx, 2015).

Our review identified 64 digital tracking measures deployed in 38 countries (see: https://bit.ly/2Zpmhgy). The textual analysis of publications on "Digital Tracking for COVID" highlights that two broad classes of technologies have been utilized to establish digital tracking:

   i. Nonmobile technologies (e.g., tracking bracelets, cameras with thermal sensors, drones), and
   ii. Mobile technologies (e.g., smartphones with inbuilt location sensors, contact-tracing mobile-based applications).

Despite the avalanche of emerging surveillance technologies to combat the pandemic, there is still a little knowledge of how these could affect society. For example, how widely the contact-tracing mobile applications are being used? What type of data will they collect, and how data will be saved? Whom is it shared with? And are there policies in place to prevent abuse?

9.3.2.1 Digital Tracking with Nonmobile Technologies

The sphere of IoT nonmobile technologies during the pandemic includes variations from the usage of electronic bracelets, deployment of cameras equipped with thermal sensors and facial recognition software, surveillance drones used to monitor mass gatherings, to extensive CCTV networks in a bid to help enforce curfews. Our review identified at least 15 of the nonmobile-based measures that were taken in 12 countries (see: https://bit.ly/2Zpmhgy). For example, in Hong Kong, all people traveling from China, Taiwan, and Macau received a bracelet to track their movements.

In West Virginia, a judge has approved strapping ankle monitors to positively tested citizens who refuse to quarantine.

In Bahrain, electronic wristbands that are compatible with the country's coronavirus contact-tracing application "BeAware" have been in use to ensure infected citizens remain quarantined. Individuals wearing the electronic bracelet must be connected to the application at all times via Bluetooth and with GPS enabled to track movement.

Corporate IoT proximity detection and contact-tracing devices have also been developed. An example comprises the LoRaWAN devices, which detect proximity and contact-tracing among employees in areas like offices and trigger alarms to alert people they should keep a safe distance.

Surveillance accompanied with facial recognition cameras equipped with enabled body temperature detection sensors has spread rapidly in China as the Chinese government is promoting the development of artificial intelligence (AI) that prioritizes the prevention of COVID-19 spread. The number of facial recognition cameras in use in China has indeed jumped 3.5 times in the past 3 years to reach 626 million deployed cameras in 2020. This motivated the government to take regulating measures to ensure the privacy of the collected biometric data through the "Personal Information Security Specifications".

Hoping to spot symptoms of the virus among the public, the authorities in Dubai are also trialing security cameras fitted with facial recognition software and thermal imaging technology. Similar thermal imaging surveillance cameras have also been tested in Bournemouth Airport, United Kingdom.

The resort city of Cannes on the Cote d'Azur has trialed innovative cameras in outdoor markets and on buses equipped with AI software that generates an automatic alert to city authorities on where breaches of the mask and distancing rules are spotted. The French firm "Datakalab" insists that its technology sets apart from the kind of hi-tech surveillance common in China as their system doesn't store any identifying data, but only sends anonymized alerts to the authorities.

Surveillance drones equipped with sensors and cameras have been also used to enforce lockdown. For example, the New York Police Department has been using aerial photography to monitor lockdown measures.

9.3.2.2   Digital Tracking with Mobile Technologies

The smartphone-enabling technologies such as Bluetooth, RFID tracking, built-in sensors, and NFC allow it to be an integral part of IoT sphere and to be the most used devices in these environments. For the digital tracking purpose during the COVID-19 crisis, authorities have strongly relied on mobile technologies. We identified 50 mobile-related digital tracking measures used in 34 countries. These can be further mapped into two classes discussed below.

9.3.2.2.1   Authorities Accessing Individual or Aggregated Data Directly from Mobile Operators

At least in 11 countries, mobile operators were reported to share individualized location data with authorities to ensure lockdown compliance. For example, in New Zealand, residents arriving from overseas were texted by the police asking if they consent to be monitored via their cell phone.

Mobile operators have been also sharing aggregated data with the authorities to facilitate the analysis of how citizens have reacted to regulations on social distancing. For instance, Vodafone in Italy reported on its website that it will be willing to assist governments in developing insights based on large, anonymized data sets by generating an aggregated heat map for the Lombardy region to "help the authorities to better understand population movements to help thwart the spread of COVID-19". A similar heatmap has been generated in Argentina as well.

9.3.2.2.2 Mobile Applications

There has been a flood of mobile applications in the middle of the battle with COVID-19. For example, in Poland, an application for quarantined patients asks to randomly take geolocated selfies. A similar application was deployed in Russia where citizens with suspected COVID-19 in Moscow were asked to send selfies three times a day to authorities. In Turkey, quarantine enforcement application, 'Life Fits Inside the House', was deployed to monitor citizens.

Wuhan, China, where the first cases of novel coronavirus were reported, opened its border with a condition imposing the installation of a government surveillance mobile application as a requirement for the entry/exit for the region. The government then tracks its citizens through the installed software by analyzing their personal data to sort individuals into color-coded categories – red, yellow, or green – corresponding to their health status and level of risk for COVID-19.

On the other hand, starting the month of April 2020, we saw an uptick in the number of a particular category of mobile applications to help stem the spread of the virus by tracking individuals and those they come into exposure with; namely "proximity contact-tracing applications". As of July 2020, there were at least 47 contact-tracing applications available for 29 countries (please see the list of the applications provided through the link: https://bit.ly/3fMGWRm). This includes the recently announced Google/Apple opt-in contact-tracing application using phone's Bluetooth connection to deliver exposure notifications for Android or iOS users (see the described scenario at Google blog: https://bit.ly/3cwZvYt).

Figure 9.1 provides the worldwide distribution for the analyzed contact-tracing applications in response to the COVID-19 pandemic along with the approximate volume of downloads as of July 2020.

While these tracing applications may have slightly different approaches on how tracing contacts, at their core, they are tracking programs using Bluetooth or GPS to track an individual's exposure to cases. Users elect to share data and are alerted if they have been within proximity to COVID-19 cases. If an individual is found to be infected with the virus, all the people that have recently been near him/her are alerted and asked to follow the public health authorities' guidelines. Not all existing contact-tracing applications serve as digital tracking applications from the authority perspective. Depending on the application design, public health authorities may (or not) receive data that users choose (or asked) to share to enhance contact tracing. Figure 9.2 provides an aggregated contemporary view of the analyzed applications.

We further analyzed the applications against the mechanisms they employ for data collection and management. Only 15% of these applications were detected collecting

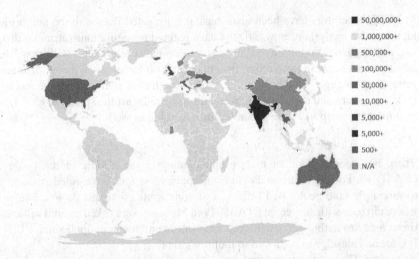

■	50,000,000+
▨	1,000,000+
■	500,000+
■	100,000+
■	50,000+
■	10,000+
■	5,000+
■	5,000+
■	500+
■	N/A

FIGURE 9.1   Distribution of COVID-19 contact tracing mobile applications as of July 2020 along with approximate volume of downloads.

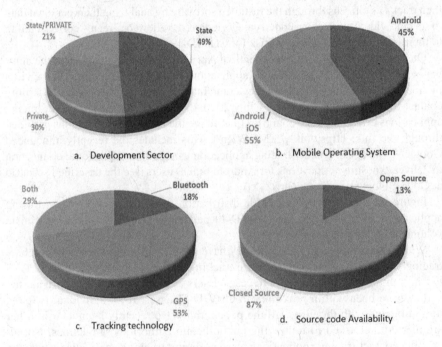

a. Development Sector

b. Mobile Operating System

c. Tracking technology

d. Source code Availability

FIGURE 9.2   Overview of the analyzed contact-tracing applications.

anonymous data with no personally identifiable information (PII). Forty-three percent of the applications maintain Pseudonymized copies of the data, while the remaining 42% of the applications had no information on the data anonymity. We report the list of the PII collected per application through the link: https://bit.ly/3fMGWRm. Once collected, 47% of the applications then stored the data in a centralized location (e.g., authority server), while 26% retained the data locally on the mobile device. Twenty-eight percent of the applications didn't report on how the data will be stored.

The majority of these applications (58%) don't provide sufficient information regarding the data storage duration though. Only 21% of these applications reported temporary storage, while the remaining 21% store collected data for a period of 1 year or longer. Forty-Seven Percent of these applications will store data in unencrypted format comparing to only 21% that store encrypted data.

Blockchain-based contact-tracing applications constitute (23%) of these applications. With the "decentralization" and "immutability" blockchain characteristics, many of the designed contact-tracing applications are utilizing a single shared ledger to store some of the identity metadata (whenever they are created), while mitigating the traditional threats to privacy due to the centralization aspects of a traditional database or cloud environment. For example, a recently constructed open-source application in India using Ethereum Blockchain allows citizens to voluntarily participate in a program that anonymously records their location history. If there are any infected people with a travel or location history originating from a particular place, the other users who were present in the close vicinity will be warned of the incident, to put them on close observation as a precaution. Storing all of the information on Ethereum is done with zero-knowledge proofs so that the identity of individuals remains pseudonymous.

While the avalanche of the digital proximity tracking technologies for COVID-19 contact tracing is understandable at the time of the pandemic, there must be ethical considerations to guide their use. The Centers for Disease Control and Prevention (CDC) defined the preliminary criteria for minimum and preferred characteristics of digital contact-tracing tools to help health departments overcome the challenges in the COVID-19 contact-tracing workflow (see: https://bit.ly/3fLAMSh). The WHO also published interim guidance in May 2020 to regulate these applications (WHO, 2020).

9.4 CONSIDERATION FOR QUALITY REQUIREMENTS FOR IoT IN HEALTHCARE APPLICATIONS

When specifying the functionality for IoT healthcare applications, attention is naturally focused on concerns such as fitness of purpose, wireless interoperability, energy efficiency, and so on. Conventional requirements elicitations techniques such as domain analysis, Joint Application Development (JAD), and Quality Function Deployment (QFD) among others (Laplante, 2013) are usually adequate for these kinds of requirements. But in healthcare IoT applications, some quality requirements are probably of greater concern. We explore these types of requirements further in this section.

9.4.1 PRIVACY REQUIREMENTS

Privacy concerns have always been a crucial aspect of healthcare. Patients expect that their PII will remain confidential and that healthcare providers will protect them. Similarly, IoT-based healthcare systems must assure privacy but allow for sharing of information that is needed to provide high-quality care across the care continuum. Many of the devices used in a provisioned, specialized IoT will collect various data whether that surveillance is known or not (Laplante, Laplante & Voas, 2015). If so, where does that data go? Who owns it? And why is it being collected in the first place? Sensors and surveillance will be huge concerns to overcome to argue convincingly for compliance when the economic benefits to healthcare providers are overwhelming for this technology.

The Health Insurance Portability and Accountability Act (HIPAA) of 1996 addresses how a patient's personal health information can be used and shared not only to assure privacy but also to allow for sharing of information that is needed to provide high-quality care. HIPAA is an integral part of today's healthcare system, and no healthcare provider would argue its relevance. New concerns, however, have been raised as to the relationship of HIPAA and the IoT.

A *Forbes* magazine report calls for new federal baseline privacy legislation, built-in security for IoT devices, data minimization (storing less, not more data), and security breach notification (McCue, 2016).

In February 2016, the Office of the Privacy Commissioner of Canada released a 34-page report (Privacy Commissioner of Canada Report, 2016) outlining concerns for privacy and data as the IoTs continues to take shape: "The Internet of Things has been compared to electricity, or a nervous system for the planet, to illustrate phenomena that are at once pervasive, unseen and will become crucially integrated within the fabric of our society", states the report. "Several international experts, thinkers and technology builders are forecasting profound political, social and economic transformations; concerns about privacy and surveillance are chief among them".

There has also been a substantial amount of academic research considering IoT functionality versus privacy. For example, Winter conducted a survey of Hawaiian healthcare consumers to identify specific "practices that will be brought about by the IoTs that may be perceived as privacy violations" (Winter, 2012). Essentially, Winter found that these consumers were willing to trade-off some privacy for the perceived benefits of information sharing. Moreover, Thierer argues against rash, restrictive regulation in response to security and privacy concerns that could thwart innovation in applications (including healthcare) related to wearable IoT technologies (Thierer, 2014). And Walla discusses some of the issues of patient privacy in the Internet-hosted personal records that are not covered by HIPAA (Walla, 2008) and Mercuri considers regulations intended to improve healthcare data access have created new security and privacy risks along with regulatory complexity for patients and practitioners (Mercuri, 2004).

9.4.1.1 Privacy Concerns with Contact-Tracing Applications

While several countries are racing to develop digital tools to improve the proximity contact tracing to control the COVID-19 virus spread, the mad dash has left some places with a confusing mishmash of options and the software security researchers' community has worried about vulnerabilities in hastily written software. For example, only 25% of the existing proximity tracing applications provide privacy policies, while only 58% do not disclose how long they will store users' data for and about 60% have no publicly stated anonymity measures.

Besides, code relating to Google advertising and tracking platforms (e.g., Google AdSense, DoubleClick) has been detected in at least 14 contact-tracing applications. We also found code relating to Facebook advertising and tracking in eight of these analyzed applications. Such a code allows publishers to make money by showing advertisements to their users from a vast array of sources. The presence of such a tracking code in contact-tracing applications raises privacy concerns due to the targeting options offered by Google/Facebook advertisement platforms.

Some of these applications are also blockchain based. Because of the blockchain's "immutability" characteristic, users' records cannot be changed or removed from the network. Will this be appropriate when a lockdown is lifted? The "finality" characteristic of the blockchain can also diverge from existing legislation such as the European General Data Protection Regulation (GDPR) or the recently announced Brazilian Lei Geral de Proteção de Dados (LGPD) that provides all citizens the capacity to govern their own data including the right to request an institution to delete personal data being processed based upon consent.

Recommending these applications is perhaps understandable in the heat of the early stages of the pandemic. However, much deliberation must be given towards considering the long-term consequences of building such digital infrastructure across society. A report published by a global public policy firm for the tech sector recalled that 49% of the world remains digitally unconnected and affirms that "virus fightback must start with adoption of policies that enable countries to take advantage of great leaps in pandemic-busting ingenuity" (D'mello, 2020; Access Partnership, 2020). The report remarks that while European countries are trying to position themselves in privacy-protecting approaches by complying with the GDPR, some Asian governments, like China and South Korea, have taken broad and aggressive approaches that rely on gathering, analyzing, and sharing vast amounts of personal data. The report also still highlights that across North and South America, there have been fewer and less-organized efforts by governments to develop contact-tracing solutions than in other regions (D'mello, 2020).

Although challenges to preserve security and privacy remain, recent studies have shown that users have felt more comfortable to use contact-tracing applications as the pandemic proceeds. For example, a survey was conducted by Metova firm, a leading provider of custom software solutions for IoT, with 2,000 residents of the United States on contact-tracing and exposure notification applications' use in the fight against COVID-19. The survey found 77% of participants would want to be notified via their mobile phone if someone they recently came in contact with was tested positive for COVID-19, and 85% are willing to anonymously share a positive COVID-19 status for the greater good (Marketing Technology Insights, 2020).

9.4.2 Safety Concerns

Safety concerns address questions such as is the system operating as intended? Is the system providing needed levels of care? Is it providing unintended functionality? Can a malfunction of the system harm a patient?

Safety requirements for medical systems often derive from oversight agencies, e.g., in the United States, the Food and Drug Administration (FDA).

Wallace and Kuhn studied 342 failures in medical devices based on data from the FDA. Their study helped identify approaches for using fault and failure information to improve device safety (Wallace & Kuhn, 2001).

They also found that "known [best] practices may not be used at all or may be misused". Wallace and Kuhn also recommended that for requirement's generation of safe medical devices that engineers should:

1. Gather failure and fault data (from previous and related systems),
2. Understand the types of faults that are prevalent for a specific domain, and
3. Develop prevention and detection approaches specific to these issues.

The US Underwriters Laboratories proposed a fault tree analysis approach for specifying hazards in wearable devices (Kirk, 2014), and this approach would be appropriate for other medical and healthcare applications using IoT technology. Using traditional techniques for defining misuse and abuse cases would also be appropriate.

9.4.3 TECHNOLOGY AND SOCIAL CONSTRAINTS

In spite the reports indicating the high level of willingness among the general population to download a contact-tracing application and share data through, the actual numbers of downloads are still relatively slim. A recent study by the University of Oxford's Big Data Institute estimates that at least 60% of the population in a given area would need to use an automated contact-tracing applications for it to be considered effective in containing the virus. On March 20, 2020, Singapore became one of the first countries to deploy a voluntary contact-tracing app, "TraceTogether" (https://www.tracetogether.gov.sg/), but only about 26% of the population installed 2 months after its inauguration. While some level of compliance is still better than none, the low rates of adoption in parts of the world are a challenge for these applications to provide any breakthrough. In addition, the lack of access to smartphones can be an obstacle to these which rely on a high level of inclusion. A median of 76% of the population across 18 advanced economies surveyed has smartphones, compared with a median of only 45% of the population in emerging economies (Silver, 2019). The dominant contact-tracing application in a nation would have to be authorized and sponsored by a government health authority. This is practically the only effective way to ensure a large-scale adoption of an application within a country. But such sponsorship process may require precious weeks for different countries' applications to be fully tested and integrated. Also, all the existing contact-tracing mobile applications rely on GPS and/or Bluetooth capabilities, which are not highly reliable still. Bluetooth's range, e.g., is considerably wider than 6 feet which can trigger a high percentage of false positives regarding the exposure to the virus. Continuously enabled GPS/Bluetooth capabilities can also drain the battery quickly. A 2016 study found that with good signal strength, a battery of a GPS-enabled mobile phone depletes by 13% while a weak signal could cause the battery to drop up to 38% (Tawalbeh & Eardley, 2016).

9.4.4 Emerging Quality Requirements for IoT in Healthcare

Caring can be described as an act or a way to approach a patient. Caring can be a trait that one possesses and often an adjective to describe what is perceived to be a "good" caregiver. Most nurses will be able to articulate their concept of caring if asked. Lachman highlights the pervasiveness of the link between nurses and caring by pointing out that "caring and nursing are so intertwined that nursing always appeared on the same page in a Google search for the definition of caring". For our purposes, we adopt caring as an adjective (functional quality) with the following definition: "displaying kindness and concern for others".

Caring likely encompasses elements of the qualities of trust, reliability, privacy, and more, but none of these, by themselves, capture the full essence of caring. Instead, caring is a super ility resulting as some composite of other ilities (and the system quality called empathy). One possible hierarchical representation for caring in terms of these other ilities is to break it down into privacy, empathy, and reliability.

Other types of systems may contain additional subqualities (e.g., usability, availability) forming a slightly different hierarchy. Since caring is comprised of some combination of other qualities that differ for each stakeholder and each system, we find it convenient to express caring for a given system as a linear combination of these constituent qualities.

"Caring" also means different things to different people and for different systems. Consider, e.g., a robotic surgery system. These systems are now used extensively for many types of procedures including heart, cancer, and prostate surgery. While current systems are robotic in the sense that the machine mimics the movements of a human surgeon, fully autonomous robot surgical systems are envisioned in the near future, replacing surgeons and nurses in the operating room (OR; Bergeles & Yang, 2014). While we expect the human surgeon and nurses to care about the patient, as systems engineers, what should we require of a fully autonomous robot surgeon? Furthermore, what should the patient expect, in terms of a caring system, especially since the patient may be unconscious during the procedure? Their concerns are likely somewhat different.

Consider, e.g., the constituent qualities of caring in the robotic surgery system. The surgeon wants the system to be safe and reliable, likely, as the primary concerns. Both the safe and trustworthy operations of the system contribute to a sense of reliability in the system and are of concern to the systems engineers. The patient shares these concerns but also wants the system to preserve his privacy (e.g., by not exposing medical records or embarrassing images). If the actors in the OR were humans, the patient would probably also expect a sense of empathy from the surgeons or nurses. Of course, robot surgeons look nothing like human surgeons; therefore, there would need to be a means by which the robots could emote empathy via speech or facial expression generation on some display device. These diverse concerns, with respect to the qualities related to caring, will inform the specific system requirements discovery and representation process.

Since many different definitions of caring exist, it is important to engage all stakeholders when trying to define a notion of caring for a new healthcare system, and it is critically important to engage systems engineers, computer scientists, doctors, nurses, and most importantly patients during requirements discovery. Many traditional requirement elicitation techniques could be used to uncover caring and related requirements depending on the size of the system. The most likely useful elicitation techniques for caring and related qualities, however, include surveys, interviews, prototyping (executable and nonexecutable), ethnographic observation, and designer as apprentice. Of course, different elicitation techniques may be used with different stakeholder groups, and multiple, complementary techniques should be used with each group.

For example, since empathy can be expressed via emoticons, prototyping (of various facial feature displays or voice outputs) could be used to generate empathy requirements. Interviews and surveys of patients could be used to capture desired caregiver behaviors (e.g., verbal cues, event triggered behaviors) that support patients in their belief that the healthcare system is trustworthy and safe. Ethnographic observation and designer as apprentice could also be used to elicit caring requirements by recording and analyzing the behaviors and movements of caregivers who are rated highly along the dimension of caring.

Of course, caring and related requirements that have been specified and delivered successfully, in-built systems could be reused in related systems and in product lines. Other caring and related requirements may emerge from laws and regulations, e.g., in the robotic surgery system case, HIPAA. Finally, other requirements for caring and related qualities will eventually emerge as standards and reference architectures are developed for applicable systems (e.g., smart healthcare).

9.5   SMART MEDICAL DEVICES

Smart medical devices are life-changing technologies similar to other medical advances such as artificial organs, prosthetics, and robotic surgery.[2] IoT has made possible the creation of devices that improve the quality of life and peace of mind of individual patients. These type of devices fall into two main categories: devices that collect data to provide information aiding in patient care and devices that are more advanced that can automate patient care. Following are some examples IoT medical devices:

- *Parkinson's Disease Movement Tracking*: Parkinson's is a brain disorder that causes physical movement problems. This watch-like movement tracking device uses a motion sensor to track abnormal movements. The patient can also use the device to record medication dosing times. The movements are recorded every 2 minutes. Comparing the movement data with the recorded medication dosing assists providers with adjusting medication decisions (Talan, 2019). Without a device like this, providers make medication adjustments based on a snapshot from a patient exam that occurs every 3 months.

[2] Some of this section has been excerpted from DeFranco, J. F., Hutchinson, M., "Understanding Smart Medical Devices," *Computer*, May 2021.

- *Smart Thermometers*: This device tracks temperature and medication times. This can provide an individual data to help determine if they are improving and also provide medication tracking – which can be difficult for someone who is ill and lives alone.
- *Smart Asthma Monitoring*: Patch devices (containing sensors) worn by the patient to detect typical asthmatic symptoms such as wheezing, respiration patterns, cough rate, heartbeat, and temperature. This can assist patients or their caregivers by providing notifications and reminders for medication dosing. With this information, a patient can also determine possible asthma attack triggers.
- *Heart Disease Wearables and Pacemakers*: Heart monitoring sensors inserted into clothing fabric. The sensors send real-time data to a server where algorithms run to determine if alerts need to be sent to a medical professional. There are also a Body Area Sensor (BAS) devices where data is used to predict cardiac arrest. This device also uses a smartphone-based heart rate detection sensor to detect a health crisis (Majumder et al., 2019). In addition, the newest generation of pacemakers use IoT architecture to monitor breathing, sinus node rate, and blood temperature. Irregularities are detected and the patient's heart rate is altered (slowing or speeding) depending on a patient's current activity level. Patient data can also be accessed through a mobile device to check device battery life and any correlations between their heart pace and activity level (Horwitz, 2019).
- *Diabetes*: As described in Chapter 1, a closed-loop insulin delivery system is an amazing device for type 1 diabetics who depend on insulin for survival. The IoT architecture consists of both a continuous glucose monitor (CGM) and an insulin pump that are worn by the patient. The CGM wirelessly sends the glucose level directly to the pump. That connection closes the loop to automatically deliver some of the insulin needed to the patient. The patient still needs to dose insulin for food consumed (insulin amount varies based on food carbohydrates.)

REFERENCES

Access Partnership, "Digital contact tracing: a comparative global study," 2020. Available at: https://www.accesspartnership.com/digital-contact-tracing-a-comparative-global-study/. Posted on May 27, 2020.

Babu, R, Jayashree, K., "A survey on the role of IoT and Cloud in health care," *International Journal of Scientific Engineering and Technology Research*, vol. 4, no. 12, 2015, pp. 2217–2219.

Bergeles, C., Yang, G. Z., "From passive tool holders to microsurgeons: safer, smaller, smarter surgical robots," *IEEE Transactions on Biomedical Engineering*, vol. 61, no. 5, 2014, pp. 1565–1576.

Caregivers & Technology: What They Want and Need, "A report published by American Association of Retired Persons," 2016. Available at: http://www.aarp.org/content/dam/aarp/home-and-family/personal-technology/2016/04/Caregivers-and-Technology-AARP.pdf.

Chouffani, R., "Can we expect the Internet of Things in healthcare?," 2016. Available at: http://internetofthingsagenda.techtarget.com/feature/Can-we-expect-the-Internet-of-Things-in-healthcare, last visited in November 2016.

Daniel, J. G., Uppaluru, M., "New reimbursement for remote patient monitoring and telemedicine," 2017. https://www.cmhealthlaw.com/2017/11/new-reimbursement-for-remotepatient-monitoring-and-telemedicine/.

D'mello, A., "Report reveals privacy is taking back seat in global contact tracing efforts," IoTNow, May 2020 [Online]. Available at: https://bit.ly/3jeKFJY.

FDA News Release, 2013. Available at: http://www.fda.gov/NewsEvents/Newsroom/PressAnnouncements/ucm369276.htm?source=govdelivery&utm_medium=email&utm_source=govdelivery.

Horwitz, J., "Medtronic debuts first apps to let heart patients monitor their pacemakers," January 16, 2019. https://venturebeat.com/2019/01/16/medtronic-debuts-first-apps-to-let-heart-patients-monitor-their-pacemakers/.

Kirk, S., "The wearables revolution: is standardization a help or a hindrance?: mainstream technology or just a passing phase?," *Consumer Electronics Magazine*, vol. 3. 2014, pp. 45–50.

Laplante, P. A., *Requirements Engineering for Software and Systems*. Taylor & Francis, Boca Raton, FL, 2013.

Laplante, P. A., Laplante, N., Voas, J., "Considerations for healthcare applications in the Internet of Things," Reliability Digest, November–December 2015. http://rs.ieee.org/images/files/techact/Reliability/2015-11/2015-11-a03.pdf.

Mack, H., "Remote patient monitoring market grew by 44 percent in 2016, report says," 2017. https://www.mobihealthnews.com/content/remotepatient-monitoring-market-grew-44-percent-2016-report-says.

Majumder, A. K. M., ElSaadany, Y., Young, R., Ucci, D., "Energy efficient wearable smart IoT System to predict cardiac arrest," *Advances in Human-Computer Interaction*, vol. 2019, 2019, pp. 1–21.

Marketing Technology Insights, "New survey reveals growing acceptance around COVID-19 con-tact tracing and exposure notification apps," June 2020. Available at: http://tiny.cc/n47asz.

Marx, G. T., "Surveillance studies," *International Encyclopedia of the Social & Behavioral Sciences*, vol. 23, no. 2, 2015, pp. 733–741.

McCue, T. J., "$117 Billion market For Internet of Things in healthcare by 2020," March 22, 2015. Available at: http://www.forbes.com/sites/tjmccue/2015/04/, last visited in November 2016.

Mercuri, R. T., "The HIPAA-potamus in health care data security," *Communications of the ACM*, vol. 47, no. 7, 2004, pp. 25–28.

National Institute of Alcohol Abuse and Alcoholism, "Treatment for alcohol problems: finding and getting help," 2016. http://pubs.niaaa.nih.gov/publications/treatment/treatment.htm, last visited in November 2016.

Patil, H. K., Seshadri, R., "Big data security and privacy issues in healthcare," *2014 IEEE International Congress on Big Data*, Anchorage, AK, pp. 762–765, 2014.

Privacy Commissioner of Canada Report, 2016. https://www.priv.gc.ca/en/, last visited in November 2016.

Qiu, Z., Gai, Z., Tao, Y., Schmitt, J., Kullak-Ublick, G. A., Wang, J., "Dual-functional plasmonic photothermal biosensors for highly accurate severe acute respiratory syndrome coronavirus 2 detection," *ACS Nano*, vol. 14, no. 5, 2020, pp. 5268–5277, pMID: 32281785. [Online]. doi:10.1021/acsnano.0c02439.

Silver, L., "Smartphone ownership is growing rapidly around the world, but not always equally," 2019. https://pewrsr.ch/2TrV7Cr.

Talan, J., "How a watch-like device is monitoring Parkinson's disease progression," Neurology Today, August 22, 2019. https://journals.lww.com/neurotodayonline/Fulltext/2019/08220/How_a_Watch_Like_Device_Is_Monitoring_Parkinson_s.8.aspx.

Tawalbeh, M., Eardley, A., "Studying the energy consumption in mobile devices," *Procedia Computer Science*, vol. 94, 2016, pp. 183–189.

Thierer, A. D., "The Internet of Things & wearable technology: addressing privacy & security concerns without derailing innovation," 2014. Available at SSRN 2494382.

Walla, K., "Redefining "user-friendliness": privacy concerns related to Non-HIPAA covered Internet Personal Health Records," Unpublished. http://law.uh.edu/healthlaw/perspectives/2008/(KW)%20PHRs.pdf.

Wallace, D., Kuhn, D. R., "Failure modes in medical device software: an analysis of 15 years of recall data," *International Journal of Reliability, Quality and Safety Engineering*, vol. 8, no. 4, 2001, pp. 351–371.

Winter, J. S., "Privacy and the emerging Internet of Things: using the framework of contextual integrity to inform policy," *Pacific Telecommunications Council Conference Proceedings*, Honolulu, Hawaii, 2012.

World Health Organization, "Ethical considerations to guide the use of digital proximity tracking technologies for COVID-19 contact tracing: interim guidance, 28 May 2020," World Health Organization, Technical Report, 2020.

10 IoT Trust Concerns[1]

10.1 INTRODUCTION

As mentioned in Chapter 2, Internet of Things (IoT) refers to systems that involve computation, sensing, communication, and actuation (NIST SP 800-183). IoT involves the connection between humans, nonhuman physical objects, and cyber objects, enabling monitoring, automation, and decision-making. The connection is complex and inherits a core set of trust concerns, most of which have no current resolution. This chapter contains NIST Internal Report 8222 (NISTIR 8222 (DRAFT)), which details 17 technical trust-related concerns for individuals and organizations before and after IoT adoption. The set of concerns discussed here is necessarily incomplete given this rapidly changing industry; however, this publication should still leave readers with a broader understanding of the topic. This set was derived from the six trustworthiness elements in NIST SP 800-183. And when possible, NISTIR 8222 (DRAFT) outlines recommendations on how to mitigate or reduce the effects of these IoT concerns. It also recommends new areas of IoT research and study.

10.2 EXECUTIVE SUMMARY OF NISTIR 8222 (DRAFT)

The IoT is used in almost every aspect of personal life and is being adopted within nearly every industry. Governments are taking notice and are looking at IoT from a variety of dimensions. One dimension is how IoT systems can improve efficiency, analytics, intelligence, and decision-making. Another dimension deals with regulation, i.e., is IoT a technology that needs governance, legislation, and standards due to its universal reach and impact? For example, IoT carries security concerns due to its high degree of connectivity. Should there be rules or laws specific to IoT security issues? And the same applies to privacy, safety, and dependability.

As with any new, unproven technology, questions about trustworthiness arise. Those questions often boil down to this: are the benefits worth the risks, i.e., are there more positive reasons to adopt a new technology than to avoid it? If answered with 'yes', a secondary question is: how can you minimize the risks to make the technology more acceptable and therefore 'suitable for use' by a wider audience? Most new technologies are created to benefit humanity; however, those technologies in the wrong hands can enable new and unforeseen nefarious actions.

NISTIR 8222 (DRAFT) is not directly focused on risk assessment and risk mitigation but instead on trust. That is, will an IoT product or service provide the desired operations with an acceptable level of quality? To answer this question, the analysis begins with a simple understanding of trust. Here, trust is the probability

---

[1] This chapter was contributed by Drs. Jeffrey Voas, Rick Kuhn, Phillip Laplante, and Sophia Applebaum.

DOI: 10.1201/9781003027799-10

that the *intended* behavior and the *actual* behavior are equivalent, given a fixed context, fixed environment, and fixed point in time. Trust is viewed as a *level of confidence*. In this publication, trust is considered at two levels: (1) can a 'thing' or device trust the data it receives, and (2) can a human trust the 'things', services, data, or complete IoT offerings that it uses. In this chapter, we are more focused on the human trust concern than the concern of 'things' to trust data (however, both are important).

NISTIR 8222 (DRAFT) promotes awareness of 17 technical concerns that can negatively affect one's ability to trust IoT products and services. It is intended for a general information technology audience including managers, supervisors, technical staff, and those involved in IoT policy decisions, governance, and procurement. This publication should be of interest to early adopters and persons responsible for integrating the various devices and services into purposed IoT offerings. The following is a brief synopsis of each technical concern.

10.2.1 SCALABILITY

This trust concern occurs from a combinatorial explosion in the number of 'things' that are part of a system. 'Things' and the services to interconnect them are often relatively inexpensive, therefore, creating an opportunity for functionality bloat. This allows complexity to skyrocket causing difficulty for testing, security, and performance. If the average person is associated with ten or more IoT 'things', the number of 'things' requiring connectivity explodes quickly and so do bandwidth and energy demands. Combinatorial explosion and functionality bloat are trust concerns.

10.2.3 HETEROGENEITY

This trust concern results from competition in the marketplace. The argument goes that with more choices, the competition will result in lower prices. While true, the ability of heterogeneous 'things' to interoperate and integrate creates a different tension related to emergent behaviors. And heterogeneity will almost definitely create *emergent behaviors* that will enable new and unknown security vulnerabilities, as well as impact other concerns such as reliability and performance. Potential vulnerability issues related to heterogeneity also occur with *supply chain* applications.

10.2.4 OWNERSHIP AND CONTROL

This trust concern occurs when much of the functionality within an IoT system originates from third-party vendors. Third-party black-box devices make trust more difficult for integrators and adopters to assess. This is particularly true for security and reliability since the internal 'workings' of black boxes are not observable and transparent. No internal computations can be specifically singled out and individually tested. Black-box 'things' can contain malicious Trojan behaviors. When IoT adopters better understand the magnitude of losing access to the internals of these acquired functions, they will recognize limitations to trust in their composite IoT systems.

10.2.5 Composability, Interoperability, Integration, and Compatibility

This trust concern occurs because hardware and software components may not work well when composed, depending on whether (1) the "right" components were selected, (2) the components had the proper security and reliability built-in, and (3) the architecture and specification of the system that the components will be incorporated into was correct. Further, problems arise if components cannot be swapped in or out to satisfy system requirements, components cannot communicate, and components cannot work in conjunction without conflict. Integration, interoperability, compatibility, and composability each impact IoT trust in a slightly different manner for networks of 'things', and each 'thing' should be evaluated before adoption into a system for each of these four properties.

10.2.6 "Ilities"

This trust concern deals with the *quality* attributes frequently referred to as "ilities". Functional requirements state what a system *shall* do. Negative requirements state what a system *shall not* do, and nonfunctional requirements, i.e., the "ilities", typically state what *level of quality* the system shall exhibit both for the functional and negative requirements. One difficulty for IoT adopters and integrators is that there are dozens of "ilities" and most are not easily measured. Another difficulty is that technically a system cannot have high levels of all "ilities" since some are in technical conflict. For example, higher security typically means lower performance. And finally, deciding which "ilities" are more important and at what level and cost is not a well understood process. No cookbook approach exists. So, although quality is desired, getting it is the challenge.

10.2.7 Synchronization

This trust concern stems from IoT systems being distributed computing systems. Distributed computing systems have different computations and events occurring concurrently. There can be numerous computations and events (e.g., data transfers) occurring in parallel, and those computations and events must need some degree of synchronization. For that to occur, a timing mechanism is needed that applies to all computations and events; however, no such global clock exists. Therefore, timing anomalies will occur, enabling vulnerabilities, poor performance, and IoT failures.

10.2.8 Measurement

This trust concern stems from a lack of IoT metrics and measures. Metrics and measures are keystones of trust. Since IoT is a relatively young set of technologies, few metrics and measures are available to adopters and integrators. To date, there are few ways to measure IoT systems other than by *counting* 'things' or dynamic testing. Because of this, it becomes difficult to argue that a system is trustable or even estimate the amount of testing that a system should receive.

10.2.9 PREDICTABILITY

This trust concern stems from an inability to predict how different components will interact. The ability to design useful IT systems depends at a fundamental level on predictability, the assurance that components will provide the resources, performance, and functions that are specified when they are needed. This is hard enough to establish in a conventional system, but an extensive body of knowledge in queuing theory and related subjects has been developed. IoT systems will provide an even greater challenge, since more components will interact in different ways, and possibly not at consistent times.

10.2.10 TESTING AND ASSURANCE

This trust concern stems from the additional testing challenges created by IoT beyond those encountered with conventional systems. The numerous number of interdependencies alone create testing difficulty because of the large numbers of tests that are needed to simply cover some percentage of the interdependencies. Testing concerns always increase when devices and services are black box and offer no transparency into their internal "workings". Most IoT systems will be built from only black-box devices and services. Also, IoT systems are highly data-driven and assuring the integrity of the data and assuring that a system is resilient to data anomalies will be required. These are just a few of the many testing and assurance problems related to IoT.

10.2.11 CERTIFICATION

This trust concern occurs because certification is difficult and often causes conflict. Questions immediately arise as to what criteria will be selected, and who will perform the certification. Other questions that arise include: (1) What is the impact on time-to-market if the system undergoes certification prior to operation? (2) What is the lifespan of a 'thing' relative to the time required to certify that 'thing'? And (3) what is the value of building a system from 'things' of which very few received certification? Without acceptable answers to such questions, it is unlikely that certification can offer the degree of trust most IoT adopters would want.

10.2.12 SECURITY

Security is a trust concern for all 'things' in IoT systems. For example, sensors data may be tampered with, stolen, deleted, dropped, or transmitted insecurely allowing it to be accessed by unauthorized parties. IoT devices may be counterfeited and default credentials are still widely used. Further, unlike traditional personal computers, there are few security upgrade processes for 'things' such as patches and updates.

10.2.13 RELIABILITY

Reliability is a trust concern for all IoT systems and 'things'. It will rarely be possible to claim that an IoT system works perfectly for any environment, context, and for any anomalous event that the system can experience. What this means for trust is that reliability assessments depend heavily on correct knowledge of the context and environment and resilience to handle anomalous events and data. Rarely will such knowledge exist and provide complete resilience.

10.2.14 DATA INTEGRITY

This trust concern focuses on the quality of the data that is generated by or fed into an IoT system. The quality of the data flowing between devices and from sensors will directly impact whether an IoT system is fit for purpose. Data is the 'blood' flowing through IoT systems. The ability to trust data involves many factors: (1) accuracy, (2) fidelity, (3) availability, (4) confidence that the data cannot be corrupted or tampered with, etc. Cloud computing epitomizes the importance of trusting data. Where data resides is important. Where is the cloud? And can the data be leaked from that location? It is a tendency to think of "your data" on "your machine". But in some cases, the data is not just "yours". Leased data can originate from anywhere and from vendors at the time of their choosing and with the integrity of their choosing. These trust concerns should be considered during IoT system development and throughout operation.

10.2.15 EXCESSIVE DATA

This trust concern is overwhelming amounts of data that gets generated and is processed in an IoT system. IoT systems are likely to have a dynamic and rapidly changing dataflow and workflow. There may be numerous inputs from a variety of sources such as sensors, external databases or clouds, and other external subsystems. The potential for the generation of vast amounts of data over time renders IoT systems as potential 'big data' generators. The possibility of not being able to guarantee the integrity of excessive amounts of data or even process that data is a trustworthiness concern.

10.2.16 PERFORMANCE

This trust concern is too much performance. This may seem counterintuitive. The speed at which computations and data generation can occur in an IoT system is increasing rapidly. Increased computational speed inhibits a system's ability to log and audit any transactions as the rate of data generation exceeds the speed of storage. This situation, in turn, makes real-time forensic analysis and recovery from faults and failures more difficult as data is lost and computational deadlines become harder to meet. Consequently, there are fewer ways to "put on the brakes", undo incorrect computations, and fix internal and external data anomalies. Furthermore, computing faster to a wrong outcome offers little trust.

10.2.17 USABILITY

This trust concern deals with whether users understand how to use the devices that that they have access to. How "friendly" are IoT devices to use and learn? This quality is an important consideration for most IT systems but may be more of a challenge with IoT, where the user interface may be tightly constrained by limited display size and functionality, or where a device can only be controlled via remote means. User interfaces for some device classes, such as smart home devices, are often limited to a small set of onboard features (e.g., LED status indicators and a few buttons) and a broader set of display and control parameters accessible remotely via a computer or mobile device. Usability and other trust concerns to which usability is intimately tied have significant implications for user trust.

10.2.18 VISIBILITY AND DISCOVERY

The visibility trust concern manifests when technologies become so ingrained into daily life that they disappear from users. If you cannot see a technology, how do you know what else it might be doing? For example, with voice response technology such as a smart speaker, when you talk to the device, do you know if it is the only system listening, and do you know if the sounds that it hears are stored somewhere for eternity and linked to you?

The discovery trust concern stems from the fact that the traditional Internet was built almost entirely on the TCP/IP protocol suite, with HTML for web sites running on top of TCP/IP. Standardized communication port numbers and internationally agreed web domain names enabled consistent operation regardless of the computer or router manufacturer. This structure has not extended to IoT devices, because they generally do not have the processing power to support it. This has enabled many new protocol families causing a vast number of possible interactions among various versions of software and hardware from many different sources. These interactions are prone to security and reliability problems.

In addition, to these the 17 concerns, this chapter concludes with two nontechnical, trust-related discussions. In the Section 10.22, the impact that many of the 17 technical concerns have on insurability and risk measurement will be discussed as well as how a lack of IoT regulatory oversight and governance affects users of IoT technologies by creating a vacuum of trust in the products and services that they can access.

10.3 INTRODUCTION OF NISTIR 8222 (DRAFT)

The IoT is being utilized in almost every aspect of life today, although this fact is often unknown and not advertised. The incorporation of IoT into everyday processes will continue to increase.

According to *Forbes* magazine (Columbus, 2017), there will be a significant increase in spending on the design and development of IoT applications and analytics. Furthermore, the biggest increases will be in the business-to-business (b2b) IoT systems (e.g. manufacturing, healthcare, agriculture, transportation, utilities), which will reach $267 billion by 2020. In addition to b2b, smart products are becoming more prevalent such as smart homes, smart cars, smart TVs, even smart light bulbs, and other basic commodities. In other words, products that can sense, learn, and react to user preferences are gaining acceptance and are deployed in modern living.

The term "Internet of Things" (IoT) is a metaphor that was coined by Kevin Ashton in 1999 (Ashton, 2009) although he prefers the phrase "Internet *for* things" (BBC, 2016). IoT is an acronym comprised of three letters: (I), (o), and (T). The (o) matters little, and as already mentioned, 'of' might be better replaced by 'for'. The Internet (I) existed long before the IoT acronym was coined, and so it is the 'things' (T) that makes IoT different from previous IT systems and computing approaches. 'Things' are what make IoT unique. Many people question whether IoT is just marketing hype or is there a science behind it. That's a fair question to ask about any new, unproven technology.

The acronym IoT currently has no universally accepted and actionable definition. However, attempts have been made. A few examples include:

- *The term Internet of Things generally refers to scenarios where network connectivity and computing capability extends to objects, sensors and everyday items not normally considered computers, allowing these devices to generate, exchange and consume data with minimal human intervention.* The Internet of Things (IoT): An Overview, Karen Rose, et.al. *The Internet Society*, October 2015. p. 5.

- *Although there is no single definition for the Internet of Things, competing visions agree that it relates to the integration of the physical world with the virtual world – with any object having the potential to be connected to the Internet via short-range wireless technologies, such as radio frequency identification (RFID), near field communication (NFC), or wireless sensor networks (WSNs). This merging of the physical and virtual worlds is intended to increase instrumentation, tracking, and measurement of both natural and social processes.* "Algorithmic Discrimination: Big Data Analytics and the Future of the Internet", Jenifer Winter. In: *The Future Internet: Alternative Visions*. Jenifer Winter and Ryota Ono, eds. Springer, December 2015. p. 127.

- *The concept of Internet of Things (IoT) ... is that every object in the Internet infrastructure is interconnected into a global dynamic expanding network.* "An efficient user authentication and key agreement scheme for heterogeneous wireless sensor network tailored for the Internet of Things environment", Mohammad Sabzinejad Farasha, et.al. *Ad Hoc Networks* 36(1), January 2016.

Instead of offering an official definition of IoT in 2016, NIST published a document titled "Networks of 'Things'" (SP 800-183 and Chapter 2 of this book) to partially address the deficit of having an accepted IoT definition (NIST, 2016). In SP 800-183, five primitives were presented that can be visualized as Lego™-like building blocks for any network of 'things'. The primitives are the (T)s.

Recall that the primitives are (1) sensors (*a physical utility that measures physical properties*), (2) aggregators (*software that transforms big data into smaller data*), (3) communication channels (*data transmission utilities that allow 'things' to communicate with 'things'*), (4) eUtilities (*software or hardware components that perform computation*), and a (5) decision trigger (*an algorithm and implementation that satisfies the purpose of a network of 'things' by creating the final output).* Note that any purposed network of 'things' may not include all five. For example, a network of 'things' can exist without sensors. And note that having a model of the components of a network of 'things' is still not a definition of IoT.

Before leaving the problem of having no universally accepted and actionable definition for IoT, it should be stated that IoT is increasingly associated with artificial intelligence (AI), automation, and 'smart' objects. So, is "IoT" any *noun* you can attach the adjective "smart" onto, e.g., smartphone, smart car, smart appliance, smart toy, smart home, smart watch, smart grid, smart city, smart tv, smart suitcase, smart clothes? No answer is offered here, but it is something to consider, because the overuse of the adjective 'smart' adds confusion as to what IoT is about.

Now consider the question: what is meant by 'trust?' No formal definition is suggested in this publication but rather a variation on the classical definition of reliability. Here, trust is the probability that the *intended* behavior and the *actual* behavior are equivalent, given a fixed context, fixed environment, and fixed point in time. Trust should be viewed as a *level of confidence.* For example, cars have a trusted set of behaviors when operating on a roadway. The same set of behaviors cannot be expected when the car is sunken in a lake. This informal trust definition works well when discussing both 'things' and networks of 'things'.

The value of knowing intended behaviors cannot be dismissed when attempting to establish trust. Lack of access to a specification for intended behaviors is a trust concern. Even if there is little difficulty gluing 'things' to other 'things', that still only addresses a network of 'things' architecture and that is one piece of determining trust. Correct architecture does not ensure that the actual behavior of the composed 'things' will exhibit the intended composite behavior. Hardware and software components may not work well when integrated, depending on whether they were the right components to be selected, whether they had the proper levels of "ilities", such as security and reliability built-in, and whether the architecture and specification for the composition was correct.

The Internet (I) is rarely associated with the term 'trust' or 'trustable'. Identity theft, false information, the dark web, breakdown in personal privacy, and other negative features of (I) have caused some people to avoid the Internet altogether. But for most, avoidance is not an option. Similar trust concerns occur for (T) because 'things' carry their own trust concerns and the interactions between 'things' can exacerbate these concerns. From a trust standpoint, the Internet should be viewed as an untrustworthy backbone with untrustworthy things attached – that becomes a perfect storm. Hence, there are three categories of IoT trust that must be addressed: (1) trust in a 'thing'; (2) trust in a network of 'things'; and (3) trust that the environment and context that the network will operate in is known and the network will be *fit for purpose* in that environment, context, and at a specific point in time.

Understanding what IoT is and what trust means is the first step in confidently relying on IoT. IoT is a complex, distributed system with temporal constraints. This publication highlights 17 technical concerns that should be considered before and after deploying IoT systems. This set has been derived from the six trustworthiness elements presented in NIST SP 800-183 (the six are reprinted in Section 10.22.)

The 17 technical concerns are (1) scalability; (2) heterogeneity; (3) control and ownership; (4) composability, interoperability, integration, and compatibility; (5) "ilities"; (6) synchronization; (7) measurement; (8) predictability; (9) IoT-specific testing and assurance approaches; (10) IoT certification criteria; (11) security; (12) reliability; (13) data integrity; (14) excessive data; (15) speed and performance; (16) usability; and (17) visibility and discovery. The publication also offers recommendations for ways to reduce the impacts of some of the 17 concerns.

This chapter also addresses two nontechnical trust concerns in Section 10.22: *insurability and risk measurement* and *lack of regulatory oversight and governance.*

In summary, NISTIR 8222 (DRAFT) advances the original six IoT trust elements presented in NIST (2016). It also serves as a roadmap for where new research and thought leadership is needed.

10.4 OVERWHELMING SCALABILITY

Computing is now embedded in products as mundane as light bulbs and kitchen faucets. When computing becomes part of the tiniest of consumer products, scalability quickly becomes an issue, particularly if these products require network connectivity. Referring back to the primitives introduced earlier, scalability issues are seen particularly with the sensor and aggregator components of IoT. Collecting and aggregating data from 10s to 100s of devices sensing their environment can quickly become a performance issue.

Consider this analysis. If the average person is associated with ten or more IoT 'things', the number of 'things' requiring connectivity explodes quickly and so do bandwidth and energy demands. Therefore, computing, architecture, and verification changes are inevitable, particularly if predictions of 20–50 billion new IoT devices being created within the next 3 years come true. More 'things' will require a means of communication between the 'things' and the consumers they serve, and the need for intercommunication between 'things' adds an additional scalability concern beyond simply counting the number of 'things' (Voas, Kuhn & Laplante, 2018a).

Increased scalability leads to increased complexity. Note that although increased scalability leads to complexity, the converse is not necessarily true. Increased complexity can arise from other factors such as infinite numbers of dataflows and workflows.

Unfortunately, complexity does not lend itself to trust that is easy to verify. Consider an analogous difficulty that occurs during software testing when the number of source lines of code (SLOC) increases. Generally, when SLOC increases, more test cases are needed to achieve greater testing coverage.[2] Simple statement testing coverage is the process of making sure that there exists a test case that touches (executes) each line of code during test. As SLOC increases, so may the number of paths though the code, and when conditional statements are considered, the number of test cases to exercise all of them thoroughly (depending on the definition of

---

[2] This difficulty does not occur for straight-line code that contains no branches or jumps, which is rare.

thoroughness) becomes combinatorically explosive.[3] IoT systems will likely suffer from a similar scalability concern that will impact their ability to have trust verified via testing.

Thus, IoT systems will likely suffer from a similar combinatorial explosion to that just mentioned for source code paths. The number of potential dataflow and workflow paths for a network of 'things' with feedback loops becomes intractable quickly, thus leading to a combinatorial explosion that impacts the ability to test with any degree of thoroughness. This is due to the expense in time and money. Further, just as occurs in software code testing, finding test scenarios to exercise many of the paths will not be feasible.[4] IoT testing concerns are discussed further in Section 10.10.

In summary, avoiding the inevitable concern of large scale for many IoT systems will not be practical. However, a network of 'things' can have bounds placed on it, e.g., limiting access to the Internet. By doing so, the threat space for a specific network of 'things' is reduced, and testing becomes more tractable and thorough. And by considering subnetworks of 'things', divide-and-conquer trust approaches can be devised that at least offer trust to higher level components than simple 'things'.

10.5   HETEROGENEITY

The heterogeneity of 'things' is economically desirable because it fosters marketplace competition. But today, IoT creates technical problems that mirror past problems when various flavors of Unix and Postscript did not interoperate, integrate, or compose well. Then, different versions of Postscript might or might not print on a specific printer and moving Unix applications to different Unix platforms did not necessarily mean the applications would execute. It was common to ask which "flavor of Unix" would a vendor's product operate on.

As with scalability, issues concerning heterogeneity are inevitable as IoT networks are developed. A network of 'things' is simply a system of 'things' that are made by various manufacturers, and these 'things' will have certain tolerances (or intolerances) to the other 'things' that they are connected to and communicate with.

The marketplace of 'things' and services (e.g., wireless communication protocols and clouds) will allow for the architecture of IoT offerings with functionality from multiple vendors. Ideally, the architecture for a network of 'things' will allow IoT products and services to be swapped in and out quickly but often that will not be the case.

Heterogeneity will create problems in getting 'things' to integrate and interoperate with other 'things', particularly when they are from different and often competing vendors, and these issues must be considered for all five classes of IoT primitives (NIST, 2016). This is discussed in Section 10.5. And heterogeneity will almost

---

[3] There are software coverage testing techniques to address testing paths and exercising complex conditional expressions; however, for these more complex forms of software testing coverage, the ability to generate appropriate test cases can become infeasible due to a lack of reachability, that is, is there any test case in the universe that can execute this scenario?

[4] This is the classic test case generation dilemma, that is, what can you do when you cannot find the type of test case you need?

definitely create *emergent behaviors* that will enable new and unknown security vulnerabilities as well as impact other concerns such as reliability and performance.

And finally, this is an appropriate place to mention potential vulnerability issues related to *supply chain*. For example, how do you know that a particular 'thing' is not counterfeit? Do you know where the 'thing' originated from? Do you trust any documentation related to the specification of a 'thing' or warranties of how the 'thing' was tested by the manufacturer? While supply chain is a concern that is too large to dwell on here with any depth, a simple principle does appear: as heterogeneity increases, it is likely that supply chain concerns will also increase.

10.6 LOSS OF OWNERSHIP AND CONTROL

Third-party black-box devices make trust more difficult for integrators and adopters to assess. This is particularly true for security and reliability in networks of 'things'. When a 'thing' is a black box, the internals of the 'thing' are not visible. No internal computations can be specifically singled out and individually tested. Black-box 'things' can contain malicious Trojan behaviors. Black boxes have no transparency.

Long-standing black-box software reliability testing approaches are a prior example of how to view this dilemma. In black-box software reliability testing, the software under test is viewed strictly by (input, output) pairs. There, the best that can be done is to build tables of (input, output) pairs, and if the tables become large enough, they can offer hints about the functionality of the box and its internals. This process becomes an informal means to attempt to reverse engineer functionality. In contrast, when source code is available, white-box testing approaches can be applied. White-box software testing offers internal visibility to the lower-level computations (e.g., at the line-of-code level).

This testing approach is particularly important for networks of 'things'. It is likely that most of the physical 'things' that will be employed in a network of 'things' will be third party, commercial, and, therefore, are commercial off-the-shelf (COTS). Therefore, visibility into the inner workings of a network of 'things' may only be possible at the communication interface layer (Voas, Charron & Miller, 1996).

Consider the following scenario: A hacked refrigerator's software interacts with an app on a person's smartphone, installing a security exploit that can be propagated to other applications with which the phone interacts. The user enters their automobile and their phone interacts with the vehicle's operator interface software, which downloads the new software, including the defect. Unfortunately, the software defect causes an interaction problem (e.g., a deadlock) that leads to a failure in the software-controlled safety system during a crash, leading to injury. A scenario such as this is sometimes referred to as a "chain of custody".

The above scenario demonstrates how losing control of the cascading events during operation can result in failure. This sequence also illustrates the challenge of identifying and mitigating interdependency risks and assigning blame when something goes wrong (using techniques such as propagation analysis and traceability analysis). And liability claims are hard to win since the "I agree to all terms" button is usually nonavoidable (Voas & Laplante, 2017). (See Section 10.13.)

Public clouds are important for implementing the economic benefits of IoT. Public clouds are black-box services. Public clouds are a commercial commodity where vendors rely on service-level agreements for legal protection from security problems and other forms of inferior service form their offerings. Integrators and adopters have few protections here. Further, what properties associated with trust can integrators and adopters test for in public clouds?

There are examples of where an organization might be able to test for some aspects of trust in a public cloud: (1) performance (i.e., latency time to retrieve data and the computational time to execute a software app or algorithm) and (2) data leakage. Performance is a more straightforward measure to assess using traditional performance testing approaches. Data leakage is harder but not impossible. By storing data that, if leaked, is easy to detect, i.e., credit card information, a bank can quickly notify a card owner when an illegitimate transaction was attempted. Note, however, such tests that do not result in the observation of leakage do not prove that a cloud is not leaking since such testing does not guarantee complete observability and is not exhaustive. This is no different than the traditional software testing problem where ten successive passing tests (meaning that no failures were observed) do not guarantee that the 11th test will also be successful.

In summary, concerns related to loss of ownership and control are often human, legal, and contractual. Technical recommendations cannot fully address these. It should be mentioned, though, that these concerns can be enumerated (e.g., as misuse or abuse cases) and evaluated during risk assessments and risk mitigation in the design and specification phases of a network of 'things'. And this risk assessment and risk mitigation may, and possibly should, continue throughout operation and deployment.

10.7 COMPOSABILITY, INTEROPERABILITY, INTEGRATION, AND COMPATIBILITY

Hardware and software components may not work well when composed, depending on whether (1) the "right" components were selected, (2) the components had the proper security and reliability built-in (as well as other quality attributes), and (3) the architecture and specification of the system that the components will be incorporated into was correct.

Note there is a subtle difference between composability, interoperability, integration, and compatibility. *Composability* addresses the issue of subsystems and components and the degree to which a subsystem or component can be swapped in or out to satisfy a system's requirements. *Interoperability* occurs at the interface level, meaning that when interfaces are understood, two distinct subsystems can communicate via a common communication format without needing knowledge concerning the functionality of the subsystems. *Integration* is a process of often bringing together disparate subsystems into a new system. And *compatibility* simply means that two subsystems can exist or work in conjunction without conflict.

Integration, interoperability, compatibility, and composability each impact IoT trust in a slightly different manner for networks of 'things', and each 'thing' should be evaluated before adoption into a system for each of these four properties.

Consider previous decades of building *systems of systems* (SoS). Engineering systems from smaller components is nothing new. This engineering principle is basic and taught in all engineering disciplines, and building networks of 'things' should be no different. However, this is where IoT's concerns of heterogeneity, scalability, and a lack of ownership and control converge to differentiate traditional SoS engineering from IoT composition.

Consider military-critical and safety-critical systems. Such systems require components that have prescriptive requirements. The systems themselves will also have prescriptive architectures that require that each component's specification is considered before adoption. Having access to information concerning the functionality, results from prior testing and expected usage of components are always required before building critical systems.

IoT systems will likely not have these prescriptive capabilities. IoT's 'things' may or may not even have specifications, and the system being built may not have a complete or formal specification. It may be more of an informal definition of what the system is to do but without an architecture for how the system should be built. Depending on (1) the grade of a system (e.g., consumer, industrial, military), (2) the criticality (e.g., safety-critical, business-critical, life-critical, security-critical), and (3) the domain (e.g., healthcare financial, agricultural, transportation, entertainment, energy), the level of effort required to specify and build an IoT system can be approximated. However, no cookbook-like guidance yet exists.

In summary, specific recommendations for addressing the inevitable issues of composability, interoperability, integration, and compatibility are (1) understand the actual behaviors of the 'things'; (2) understand the environment, context, and timing that each 'thing' will operate in, (3) understand the communication channels between the 'things' (NIST, 2016), (4) apply systems of systems design and architecture principles when applicable, (5) and apply the appropriate risk assessment and risk mitigation approaches during architecture and design based on the grade, criticality, and domain.

10.8 ABUNDANCE OF "ILITIES"

A trust concern for networks of 'things' deals with the *quality* attributes termed "ilities" (Voas, 2004). Functional requirements state what a system *shall* do. Negative requirements state what a system *shall not* do, and nonfunctional requirements, i.e., the "ilities" typically state what *level of quality* the system shall exhibit both for the functional and negative requirements. "Ilities" apply to both 'things' and the systems they are built into.

It is unclear how many "ilities" there are – it depends on who you ask. This document mentions each of these "ilities" in various contexts and level of detail: availability, composability, compatibility, dependability, discoverability, durability, fault tolerance, flexibility, interoperability, insurability, liability, maintainability, observability, privacy, performance, portability, predictability, probability of failure, readability, reliability, resilience, reachability, safety, scalability, security, sustainability, testability, traceability, usability, visibility, vulnerability. Most of these will apply to 'things' and networks of 'things'. However, not all readers will consider all of these to be legitimate "ilities".

One difficulty here is that for some "ilities", there is a subsumes hierarchy. For example, reliability, security, privacy, performance, and resilience are "ilities" that are grouped into what LaPrie et Al. termed as "dependability"[5]. While having a subsumes hierarchy might appear to simply the relationship between different "ilities", that is not necessarily the case. This can create confusion.

Building levels of the "ilities" into a network of 'things' is costly and not all "ilities" cooperate with each other, i.e., "building in" more security can reduce performance (Voas & Hurlburt, 2015). Another example would be fault tolerance and testability. Fault-tolerant systems are designed to mask errors during operation. Testable systems are those that do not mask errors and make it easier for a test case to notify when something is in error inside of a system. Deciding which "ilities" are more important is difficult from both a cost-benefit trade-off analysis and a technical trade-off analysis. Also, some "ilities" can be quantified and others cannot. For those that cannot be quantified, qualified measures exist.

Further, consider an "ility" such as reliability. Reliability can be assessed for (1) a 'thing', (2) the interfaces between 'things', and (3) the network of 'things itself (Voas, 1997). And these three types of assessments apply to most "ilities".

Deciding which "ilities" are more important and at what level and cost is not a well-understood process. No cookbook approach exists. The point here is that these nonfunctional requirements often play just as important of a role in terms of the overall system quality as do functional requirements. This reality will impact the satisfaction of the integrators and adopters with the resulting network.

In summary, deciding which "ility" is more important than others must be dealt with on a case–by-case basis. It is recommended that the "ilities" are considered at the beginning of the life cycle of a network of 'things'. Failure to do so will cause downstream problems throughout the system's life cycle, and it may continually cause contention as to why intended behaviors do not match actual behaviors.

10.9 SYNCHRONIZATION

A network of 'things' is a distributed computing system. Distributed computing systems have different computations and events occurring concurrently. There can be numerous computations and events (e.g., data transfers) occurring in parallel.

This creates an interesting dilemma, similar to that in air traffic control: trying to keep all events properly synchronized and executing at the precise times and in a precise order. When events and computations get out of order due to delays or failures, an entire ecosystem can become unbalanced and unstable.

IoT is no different and possibly more complex than air traffic control. In air traffic control, there is a basic global clock that does not require events be time-stamped to

---

[5] From Wikipedia: In systems engineering, **dependability** is a measure of a system's **availability**, **reliability**, and its **maintainability**, and **maintenance support performance**, and, in some cases, other characteristics such as **durability**, **safety** and **security** (1). In software engineering, **dependability** is the ability to provide services that can defensibly be trusted within a time-period (2). This may also encompass mechanisms designed to increase and maintain the dependability of a system or software (3).

high levels of fidelity, e.g., a microsecond. Further, events are regionalized around particular airspace sectors and airports.

There is nothing similar in IoT. Events and computations can occur anywhere, be transferred at "any time", and occur at differing levels of speed and performance. The desired result is that all these events and computations converge towards a single decision (output). The key concern is "any time", because these transactions can take place geographically anywhere, at the microsecond level, and with no clear understanding of what the clock in one geographic region means with respect to the clock in another geographic region.

There is no trusted universal time-stamping mechanism for practical use in many or most IoT applications. The Global Positioning System (GPS) can provide very precise time, accurate up to 100 ns with most devices. Unfortunately, GPS devices have two formidable limitations for use in IoT. First, GPS requires unobstructed line-of-sight access to satellite signals. Many IoT devices are designed to work where a GPS receiver could not receive a signal, such as indoors or otherwise enclosed in walls or other obstructions. Additionally, even if an IoT device is placed where satellite signal reception is available, GPS power demands are significant. Many IoT devices have drastically limited battery life or power access, requiring carefully planned communication schedules to minimize power usage. Adding the comparatively high-power demands of GPS devices to such a system could cripple it, so in general GPS may not be practical for use in many networks of things.

Consider a scenario where a sensor in geographic location v is supposed to release data at time x. There is an aggregator in location z waiting to receive this sensor's data concurrently with outputs from other sensors. Note that v and z are geographically far apart and the local time x in location v does not agree, at a global level, with what time it is at z. If there existed a universal time-stamping mechanism, local clocks could be avoided altogether, and this problem would go away. With universal time-stamping, the time of every event and computation in a network of 'things' could be agreed upon by using a central time-stamping authority that would produce time stamps for all events and computations that request them. Because timing is a vital component needed to trust distributed computations, such an authority would be beneficial. However, such an authority does not exist (Stavrou & Voas, 2017). Research is warranted here.

10.10 LACK OF MEASUREMENT

Standards are intended to offer levels of trust, comparisons of commonality, and predictions of certainty. Standards are needed for nearly everything, but without metrics and measures, standards become more difficult to write and determine compliance against. Metrics and measures are classified in many ways.

Measurement generally allows for determination of one of two things: (1) what currently exists, and (2) what is predicted and expected in the future. The first is generally easier to measure. One example is *counting*. For example, one can count the number of coffee beans in a bag. Another approach is estimation. *Estimation* approximates what you have. By using the coffee example and having millions of beans to count, it might be easier weighing the beans and using that weight to estimate an approximate count.

Prediction is different than estimation, although estimation can be used for prediction. For example, an estimate of the current reliability of a system, given a fixed environment, context, and point in time might be 99%. Note the key word is point in time. In comparison, a prediction would say something like: based on an estimate of 99% reliability today, it is believed that the reliability will also be 99% reliable tomorrow, but after tomorrow, the reliability might change. Why? The reason is simple: As *time* moves forward, components usually wear out, thus reducing overall system reliability. Or as time moves forward, the environment may change such that the system is under less stress, thus increasing predicted reliability. In IoT, as 'things' may be swapped in and out on a quick and continual basis, predictions and estimations of an "ility" such as reliability will be difficult.

To date, there are few ways to measure IoT systems other than by *counting* 'things' or dynamic testing. Counting is a static approach. Testing is a dynamic approach when the network is executed. (Note that there are static testing approaches that do not require network execution, e.g., a walkthrough of the network architecture.) Thus, the number of 'things' in a system can be counted just like how lines of code in software can be counted, and black-box testing can be used to measure certain "ilities".

In summary, several limited recommendations have been mentioned for mitigating the current lack of measurement and metrics for IoT. To date, counting measures and dynamic approaches such as estimating reliability and performance are reasonable candidates. Static testing (e.g., code checking) can also be used to show that certain classes of IoT vulnerabilities are likely not present. IoT metrology is an open research question.

10.11 PREDICTABILITY

The ability to design useful IT systems depends at a fundamental level on predictability, the assurance that components will provide the resources, performance, and functions that are specified when they are needed. This is hard enough to establish in a conventional system, but an extensive body of knowledge in queuing theory and related subjects has been developed. IoT systems will provide an even greater challenge, since more components will interact in different ways and possibly not at consistent times.

Two properties of IoT networks have a major impact on predictability: (1) a much larger set of communication protocols may be involved in a single network, and (2) the network configuration changes rapidly. Communication protocols for networks of 'things' include at least 13 data links, 3 network layer routings, 5 network layer encapsulations, 6 session layers, and 2 management standards (Salman). Data aggregators in the network must thus be able to communicate with devices that have widely varying latency, throughput, and storage characteristics. Since many small devices have limited battery life, data transmission times must be rationed, so devices are not always online. For example, Bluetooth Low Energy (BLE) devices can be configured to broadcast their presence for periods ranging from 0.2 to 10.2 seconds.

In addition to second-by-second changes in the set of devices currently active, another issue with network configuration changes stems from the embedding of computing devices with the physical world. Even more than conventional systems, humans are part of IoT systems, and necessarily affect the predictable availability of services, often in unexpected ways. Consider the story of a driver who took advantage of a cell phone app that interacts with his vehicle's onboard network to allow starting the car with the phone. Though probably not considered by the user, the starting instructions are routed through the cellular network. The car owner started his car with the cell phone app, then later parked the car in a mountainous area, only to discover that it was impossible to restart the car because there was no cell signal (Neumann, 2018).

This rather amusing story illustrates a basic predictability problem for IoT networks – node location and signal strength may be constantly changing. How do you know if a constantly changing network will continue to function adequately and remain safe? Properties such as performance and capacity are unavoidably affected as the configuration evolves, but you need to be able to predict these to know if and how a system can be used for specific purposes. Modeling and simulation become essential for understanding system behavior in a changing environment, but trusting a model requires some assurance that it incorporates all features of interest and accurately represents the environment. Beyond this, it must be possible to adequately analyze system interactions with the physical world, including potentially rare combinations of events.

Recommendations for design principles will evolve for this new environment but will take time before users are able to trust systems composed often casually from assorted components. Here again, the importance of a central theme of this document is reshown: to be able to trust a system, it must be bounded, but IoT by its nature may defy any ability to bound the problem.

10.12 FEW IoT-SPECIFIC TESTING AND ASSURANCE APPROACHES

To have any trust in networks of 'things' acting together, assurance will need to be much better than it is today. A network of 'things' presents a number of testing challenges beyond those encountered with conventional systems. Some of the more significant include the following:

- *Communication among Large Numbers of Devices*: Conventional Internet-based systems usually include one or more servers responding to short communications from users. There may be thousands of users, but the communication is typically one to one, with possibly a few servers cooperating to produce a response to users. Networks of 'things' may have several tens to hundreds of devices communicating.
- *Significant Latency and Asynchrony*: Low power devices may conserve power by communicating only on a periodic basis, and it may not be possible to synchronize communications.
- *More Sources of Failure*: Inexpensive, low power devices may be more likely to fail, and interoperability problems may also occur among devices with slightly different protocol implementations. Since the devices may have limited storage and processing power, software errors in memory management or timing may be more common.

- *Dependencies among Devices Matter*: With multiple nodes involved in decisions or actions, some nodes will typically require data from multiple sensors or aggregators, and there may be dependencies in the order this data is sent and received. The odds of failure increase rapidly as the chain of cooperating devices grows longer.

The concerns listed above produce a complex problem for testing and assurance, exacerbated by the fact that many IoT applications may be safety critical. In these cases, the testing problem is harder, but the stakes may be higher than for most testing. For essential or life-critical applications, conventional testing and assurance will not be acceptable.

For a hypothetical example, consider a future remote health monitoring and diagnosis app, with four sensors connected to two aggregators, which are connected to an *e*Utility that is then connected to a local communication channel, which in turn connects to the external Internet, and finally with a large artificial intelligence (AI) application at a central decision trigger node. While 99.9% reliability might seem acceptable for a $3.00 device, it will not be, if included in a critical system. If correct operation depends on all ten of these nodes, and each node is 99.9% reliable, then there is nearly a 1% chance that this network of things will fail its mission, an unacceptable risk for life-critical systems. Worse, this analysis has not even considered the reverse path from the central node with instructions back to the originating app.

Basic recommendations to reduce this level of risk include redundancy among nodes, and much better testing. This means not just more of conventional test and review activities but different kinds of testing and verification. For some IoT applications, it will be necessary to meet test criteria closer to what are used in applications such as telecommunications and avionics, which are designed to meet requirements for failure probabilities of 10^{-5} and 10^{-9}, respectively. Redundancy is part of the answer, with a trade-off that interactions among redundant nodes become more critical and the redundant node interactions are added to the already large number of interacting IoT nodes.

One additional testing and assurance issue concerns the *testability* of IoT systems (Voas, Kuhn & Laplante, 2018b). There are various meanings of this "ility"; however, two that apply here are (1) the ability of testing to detect defects, and (2) the ability of testing to cover[6] (execute) portions of the system using a fixed set of test cases. The reason (1) is a concern that IoT systems may have small output ranges, e.g., a system may only produce a binary output. Such systems, if very complex, may inherit an ability to hide defects during testing. The reason (2) is a concern that if high levels of test coverage cannot be achieved, more portions of the overall system will go untested leaving no clue as to what might happen when those portions are executed during operation.

The key problem for IoT testing is apparent from the test issues discussed above – huge numbers of interactions among devices and connections, coupled with order dependencies. Fortunately, methods based on combinatorics and design of

---

[6] Coverage too comes in different types, for instance, the ability to execute each 'thing' once is different than executing each path through a system once.

experiments work extremely well in testing complex interactions (Patil, Goveas & Rangarajan, 2015; Dhadyalla, Kumari & Snell, 2014; Yang, Zhang & Fu, 2013). Covering array generation algorithms compress huge numbers of input value combinations into arrays that are practical for most testing, making the problem more tractable, and coverage more thorough, than would be possible with traditional use case-based testing. Methods of dealing with this level of testing complexity are the subject of active research (Voas, Kuhn & Laplante, 2018b).

10.13   LACK OF IoT CERTIFICATION CRITERIA

Certification of a product (not processes or people) is a challenge for any hardware, software, service, and hybrid systems (Voas, 1998a,b 1999, 2000; Voas & Payne, 2000; Miller & Voas, 2006). IoT systems are hybrids that may include services (e.g., clouds) along with hardware and software.

If rigorous IoT certification approaches are eventually developed, they should reduce many of the trust concerns in this publication. However, building certification approaches is generally difficult (Voas, 1999). One reason is that certification approaches have less efficacy unless correct threat spaces and operational environments are known. Often, these are not known for traditional systems, let alone for IoT systems.

Certification economics should also be considered, e.g., the cost to certify a 'thing' relative to the value of that 'thing'. The *criteria* used during certification must be rigorous enough to be of value. And a question of who performs the certification and what their qualifications are to perform this work cannot be overlooked. Two other considerations are: (1) what is the impact on the time to market of a 'thing' or network of 'things'? And (2) what is the lifespan of a 'thing' or network of 'things'? These temporal questions are important because networks of 'things' along with their components may have short lives that far exceed the time needed to certify.

Certifying 'things' as standalone entities does not solve the problem of system trust, particularly for systems that operate in a world where their environment and threat space is in continual flux.

If 'things' have their functional and nonfunctional requirements defined, they can be vetted to assess their ability to: (1) be integrated, (2) communicate with other 'things', (3) not create conflict (e.g., no malicious output behaviors), and (4) be swapped in and out of a network of 'things' (e.g., when a newer or replacement 'thing' becomes available).

When composing 'things' into systems, special consideration must be given if all of the 'things' are not certified. For example, not all 'things' in a system may have equal significance to the functionality of the system. It would make sense to spend vetting resources on those that have the greatest impact. Therefore, weighting the importance of each 'thing' should be considered and then decide what to certify and what to ignore based on the weightings. And if all 'things' are certified, that still does not mean they will interoperate correctly in a system because the environment, context, and threat space all play a key role in that determination.

And perhaps most importantly, what functional, nonfunctional, or negative behavior is being certified for? And are forms of vetting available to do that? For example, how can a network of 'things' demonstrate that certain security vulnerabilities are not present?

In summary, limited recommendations can be considered for how to certify 'things' and systems of 'things'. Software testing is a first line of defense for performing lower levels of certification; however, it is costly and can overestimate quality, e.g., you test a system twice and if it works, potentially leading to a false assumption that the system is reliable and does not need a third test. Probably a good first step here is to first define the type of quality you are concerned about. (See Section 10.6.) From there, you can assess what can be certified in a timely manner and at what cost.

10.14   SECURITY

Like traditional IT or enterprise security, IoT security is not a one-size-fits-all problem, and the solutions deployed to this problem tend to only be quick fixes that push the issue down the line. Instead, it should be recognized that the issue of IoT security is both multifaceted and dependent on the effort to standardize IoT security. This section walks through several of these important facets, highlighting solutions that do exist and problems that remain to be solved.

10.14.1   SECURITY OF 'THINGS'

Security is a concern for all 'things'. For example, sensors and their data may be tampered with, stolen, deleted, dropped, or transmitted insecurely allowing it to be accessed by unauthorized parties. Further, sensors may return no data, totally flawed data, or partially flawed data due to malicious intent. Sensors may fail completely or intermittently and may lose sensitivity or calibration due to malicious tampering. Note however that building security into specific sensors may not be cost-effective depending on the value of a sensor or the importance of the data it collects. Aggregators may contain malware affecting the correctness of their aggregated data. Further, aggregators could be attacked, e.g., by denying them the ability to execute or by feeding them bogus data. Communication channels are prone to malicious disturbances and interruptions.

The existence of counterfeit 'things' in the marketplace cannot be dismissed. Unique identifiers for every 'thing' would be ideal for mitigating this problem, but that is not practical. Unique identifiers can partially mitigate this problem by attaching RFID tags to physical primitives. RFID readers that work on the same protocol as the inlay may be distributed at key points throughout a network of 'things'. Readers activate a tag causing it to broadcast radio waves within bandwidths reserved for RFID usage by individual governments internationally. These radio waves transmit identifiers or codes that reference unique information associated with the item to which the RFID inlay is attached, and in this case, the item would be a physical IoT primitive.

The time at which computations and other events occur may also be tampered with, making it unclear when events actually occurred, not by changing time (which is not possible), but by changing the recorded time at which an event in the workflow is generated, or computation is performed, e.g., sticking in a **delay**() function call.

Malicious latency to induce delays are possible and will affect when decision triggers are able to execute.

Thus, networks of 'things', timing, and 'things' themselves are all vulnerable to malicious intent.

10.14.2 Passwords

Default credentials have been a problem plaguing the security community for some time. Despite the many guides that recommend users and administrators change passwords during system setup, IoT devices are not designed with this standard practice in mind. In fact, most IoT devices often lack intuitive user interfaces with which credentials can be changed. While some IoT device passwords are documented either in user manuals or on manufacturer websites, some device passwords are never documented and are unchangeable. Indeed, both scenarios can be leveraged by botnets. The Mirai botnet and its variants successfully brute forced IoT device default passwords to ultimately launch distributed denial-of-service attacks against various targets (Kolias et al., 2017).

Many practitioners have proposed solutions to the problem of default credentials in IoT systems, ranging from the usual recommendation to change credentials – perhaps with more user awareness – to more advanced ideas like encouraging manufacturers to randomize passwords per device. While not explicitly mitigating the problem of default credentials, the Manufacturer Usage Description (MUD) specification (Lear, Droms & Romascanu, 2017) allows manufacturers to specify authorized network traffic, which can reduce the damage caused by default credentials. This specification employs a defense-in-depth strategy intended to address a variety of problems associated with the widespread use of sensor-enabled end devices such as IP cameras and smart thermostats. MUD reduces the threat surface of an IoT device by explicitly restricting communications to and from the IoT device to sources and destinations intended by the manufacturer. This approach prevents vulnerable or insecure devices from being exploited and helps alleviate some of the fallout of manufacturers leaving in default credentials.

10.14.3 Secure Upgrade Process

On a traditional personal computer, weaknesses are typically mitigated with patches and upgrades to various software components, including the operating system. On established systems, these updates are usually delivered via a secure process, where the computer can authenticate the source pushing the patch. While parallels exist for IoT devices, very few manufacturers have secure upgrade processes with which to deliver patches and updates; oftentimes, attackers can man-in-the-middle the traffic to push their own malicious updates to the devices, thereby compromising them. Similarly, IoT devices can receive feature and configuration updates, which can likewise be hijacked by attackers for malicious effect.

Transport standards such as HTTPS as well as existing public key infrastructure provide protections against many of the attacks that could be launched against upgrading IoT devices. These standards, however, are agnostic on the implementations of the IoT architecture and do not cover all the edge cases. However, the IoT Firmware Update Architecture (Moran, Meriac & Tschofenig, 2017) – recently proposed to the IETF – provides necessary details needed to implement a secure firmware update architecture, including hard rules defining how device manufacturers should operate. Following this, emerging standard could easily mitigate many potential attack vectors targeting IoT devices.

10.14.4 SUMMARY

Addressing the security of IoT devices is a prescient issue as IoT continues to expand into daily life. While security issues are widespread in IoT ecosystems, existing solutions – such as MUD to remediate password weaknesses and transport standards for secure upgrades – can be leveraged to boost the overall security of devices. Deploying these existing solutions can yield significant impacts on the overall security without requiring significant amounts of time spent researching new technologies.

10.15 RELIABILITY

IoT reliability should be based on the traditional definition in Musa, Iannino and Okumoto (1987). The traditional definition is simply the probability of failure-free operation of individual components, groups of components, or the whole system over a bounded time interval and in a fixed environment. Note that is what the informal definition of trust mentioned earlier was based on. This definition assumes a static IoT system, meaning new 'things' are not continually being swapped in and out. But realistically, that will not be the case since new 'things' will be added dynamically and on the fly, either deliberately or inadvertently. Thus, the instantaneously changing nature of IoT systems will induce emergent, and complex chains of custody make it difficult to insure and correctly measure reliability (Miller, Voas & Laplante, 2010; Voas, Kuhn & Laplante, 2018a). The dynamic quality of IoT systems requires that reliability be reassessed when components change and the operating environment changes.

Reliability is a function of context and environment. Therefore, to perform reliability assessments, a priori knowledge of the appropriate environment and context is needed. It will rarely be possible to make a claim such as *this network of 'things' works perfectly for any environment, context, and for any anomalous event that the system can experience.* Unfortunately, wrong assumptions about environment and context will result in wrong assumptions about the degree to which trust has been achieved.

To help distinguish the difference between context and environment, consider a car that fails after a driver breaks an engine by speeding above the manufacturer's maximum expectation while driving in excellent road conditions and good weather. Weather and road conditions are the environment. Speeding past the manufacturer's maximum expectation is the context. Violating the expected context or expected environment can both impact failure. But here, failure occurred due to context.

The relationship between anomalous events and 'things' is important for a variety of reasons, not the least of which is the loss of ownership and control already mentioned. Assume worst-case scenarios from 'things' that are complete black boxes.

Consider certain scenarios: (1) a 'thing' fails completely or in a manner that creates bad data that infects the rest of the system, and (2) a 'thing' is fed corrupt data and you wish to know how that 'thing' reacts, i.e., is it resilient? Here, resilience means that the 'thing' still provides acceptable behavior. These two scenarios have been referred to as "propagation across" and "propagation from" (Voas, 1997). Propagation across is the study of "garbage in garbage out". Propagation across tests the strength of a component or 'thing'. Propagation from is the study of how far through a system an internal failure that creates corrupt data can cascade. Possibly it propagates all of the way and the system fails, or possibly the corrupted internal state of the system is not severe enough to cause that. In this case, the system shows its resilience.

A related concern involves who is to blame when a 'thing' or network of 'things' fails? This trust concern (and legal liability) becomes especially problematic when there are unplanned interactions between critical and noncritical components. In discussing IoT trust, there are two related questions: (1) What is the possibility of system failure? And (2) who is liable when the system fails? (Voas & Laplante, 2017).

Consider the first question: What is the possibility of system failure? The answer to this question is very difficult to determine. A powerful technique for determining the risks of a system-level failure would involve fault injection to simulate the effects of real faults as opposed to simulating the faults themselves. But until these risks can be accurately and scientifically measured, there likely won't be a means for probabilistically and mathematically bounding and quantifying liability (Voas & Laplante, 2017).

Now consider the second question: Who is liable when the system fails? For any noninterconnected system, the responsibility for failure lies with the developer (i.e., the individual, individuals, company, or companies, inclusive). But for systems that are connected to other systems locally and through the Internet, the answer becomes more difficult. Consider the following legal opinion:

> In case of (planned) interconnected technologies, when there is a 'malfunctioning thing' it is difficult to determine the perimeter of the liability of each supplier. The issue is even more complex for artificial intelligence systems involving a massive amount of collected data so that it might be quite hard to determine the reason why the system made a specific decision at a specific time. (Coraggio, 2016)

Interactions, both planned and spontaneous, between critical and noncritical systems create significant risk and liability concerns. These interacting, dynamic, cross-domain ecosystems create the potential for increased threat vectors, new vulnerabilities, and new risks. Unfortunately, many of these will remain as unknown unknowns until after a failure or successful attack has occurred.

In summary, this publication offers no unique recommendations for assessing and measuring reliability. The traditional reliability measurement approaches that have been around for decades are appropriate for a 'thing' and a network of 'things'. These approaches, as well as assessments of resilience, should be considered throughout a system's life cycle.

10.16 DATA INTEGRITY

Data is the 'blood' of any computing system including IoT systems. And if a network of 'things' involves many sensors, there may be a lot of data.

The ability to trust data involves many factors: (1) accuracy, (2) fidelity, (3) availability, (4) confidence that the data cannot be corrupted or tampered with, etc. Whether any of these is more important than the other depends on the system's requirements; however, with respect to a network of 'things', the timeliness with which the data is transferred is of particular importance. Stale, latent, and tardy data is a trust concern, and while that is not a direct problem with the "goodness" of the data itself, it is a performance concern for the mechanisms within the network of 'things' that transfer data. In short, stale, latent, and tardy data in certain situations will be no worse than no data at all.

Cloud computing epitomizes the importance of trusting data. Where data resides is important. Where is the cloud? And can the data be leaked from that location? It is a tendency to think of "your data" on "your machine". But in some cases, the data is not just "yours". Leased data can originate from anywhere and from vendors at the time of their choosing and with the integrity of their choosing. Competitors can lease the same data (Miller, Voas & Laplante, 2010; NIST, 2016).

The production, communication, transformation, and output of large amounts of data in networks of 'things' create various concerns related to trust. A few of these include:

1. *Missing or Incomplete Data*: How does one identify and address missing or incomplete data? Here, missing or incomplete data could originate from a variety of causes, but in IoT, it probably refers to sensor data that is not released and transferred or databases of information that are inaccessible (e.g., clouds). Each network of 'things' will need some level of resilience to be built-in to allow a potentially crippled network of 'things' to still perform even when data is missing or incomplete.

2. *Data Quality*: How does one address data quality? To begin, a definition is needed for what data quality means for a particular system. Is it fidelity of the information, accuracy of the information, etc.? Each network of 'things' will need some description for what an acceptable level of data quality is.

3. *Faulty Interfaces and Communication Protocols*: How does one identify and address faulty interfaces and communication protocols? Here, since data is the 'blood' of a network of 'things', the interfaces and communication protocols are the veins and arteries of that system. Defective mechanisms that perform data transfer within a system if 'things' are equally as damaging to the overall trust in the data as is poor data quality, missing data, and incomplete data. Therefore, trust must exist in the data transfer mechanisms. Each network of 'things' will need some level of resilience to be built in to ensure that the data moves from point A to point B in a timely manner. This solution might include fault-tolerance techniques such as redundancy of the interfaces and protocols.

4. *Data Tampering*: How does one address data tampering or even know it occurred? Rarely can tamperproof data exist if someone has malicious intent and the appropriate resources to fulfill that intent. Each network of 'things' will need some type of a reliance plan for data tampering, such as a back-up collection of the original data and in a different geographic location.

5. *Data Security and Privacy*: How secure and private is the data from delay or theft? There are a seemingly infinite number of places in the dataflow of a network of 'things' where data can be snooped by adversaries. This requires that the specifications of a network of 'things' have had some risk assessment that assigns weights to the value of the data if it were to be compromised. Each network of 'things' will need a data security and privacy plan.

6. *Data Leakage*: Can data leak, and if so, would you know that it had? Assume worst-case scenario where all networks of 'things' leak. While this does not directly impact the data, it may well impact the business model of the organization that relies on the system of 'things'. If this is problematic, an analysis of where the leakage could originate can be performed; however, this is technically difficult and costly.

While conventional techniques such as error correcting codes, voting schemes, and Kalman filters could be used, specific recommendations for design principles need to be determined on a case-by-case basis.

10.17   EXCESSIVE DATA

Any network of 'things' is likely to have a dynamic and rapidly changing dataflow and workflow. There may be numerous inputs from a variety of sources such as sensors, external databases or clouds, and other external subsystems. The potential for the generation of vast amounts of data over time renders IoT systems as potential 'big data' generators. In fact, one report predicts that global data will reach 44 zettabytes (44 billion terabytes) by 2020 (Data IQ News, 2014). (Note however there will be networks of 'things' that are not involved in receiving or generating large quantities of data, e.g., closed-loop systems that have a small and specialized purpose. An example here would be a classified network that is not tethered to the Internet.)

The data generated in any IoT system can be corrupted by sensors, aggregators, communications channels, and other hardware and software utilities (NIST, 2016). Data is not only susceptible to accidental corruption and delay but also malicious tampering, delay, and theft. As previously mentioned in Section 10.14, data is often the most important asset to be protected from a cybersecurity perspective.

Each of the primitives presented in NIST (2016) is a potential source for a variety of classes of corrupt data. Section 10.13 already discussed the problems of "propagation across" and "propagation from". Although hyperbole, it is reasonable to visualize an executing network of 'things' to a firework show. Different explosions occur at different times although all are in timing coordination during a show. Networks of 'things' are similar in that internal computations, and the resulting data is in continuous generation until the IoT system performs an actuation or decision.

The dynamic of data being created quickly and used to create new data and so on cannot be dismissed as a problem for testing and any hope of traceability and observability when an unexpected behavior occurs. Thus, the vast amount of data that can be generated by networks of 'things' makes the problem of isolating and treating corrupt data extremely difficult. The difficulty pertains to the problem of identifying corrupt data and the problem of making this identification quickly enough. If such identification cannot be made for a certain system in a timely manner, then trust in that system is an unreasonable expectation (Voas, Kuhn & Laplante, 2018b).

Certain data compression, error detection and correction, cleaning, filtering, and compression techniques may be useful both in increasing trust in the data and reducing its bulk for transmission and storage. No specific recommendations, however, are made.

10.18 SPEED AND PERFORMANCE

The speed at which computations and data generation can occur in a network of 'things' is increasing rapidly. Increased computational speed inhibits a system's ability to log and audit any transactions as the rate of data generation exceeds the speed of storage. This situation, in turn, makes real-time forensic analysis and recovery from faults and failures more difficult as data is lost and computational deadlines become harder to meet. Consequently, there are fewer ways to "put on the brakes", undo incorrect computations, and fix internal and external data anomalies. Furthermore, computing faster to a wrong outcome offers little trust.

A related problem is that of measuring the speed of any network of 'things'. Speed-oriented metrics are needed for optimization, comparison between networks of 'things', and identification of slowdowns that could be due to anomalies – all of which affect trust.

But there are no simple speed metrics for IoT systems and no dashboards, rules for interoperability and composability, rules of trust, and established approaches to testing (Voas, Kuhn & Laplante, 2018a).

Possible candidate metrics to measure speed in an IoT system include:

1. Time to decision once all requisite data is presented; this is an end-to-end measure.
2. Throughput speed of the underlying network,
3. Weighted average of a cluster of sensor's "time to release data",
4. Some linear combination of the above or other application domain specific metrics.

Note here that while better performance will usually be an "ility" of desire, it makes the ability to perform forensics on system that fail much harder, particularly, for systems where some computations occur so instantaneously that there is no "after the fact" trace of them.

Traditional definitions from real-time systems engineering can also be used, for example:

1. *Response Time*: The time between the presentation of a set of inputs to a system and the realization of the required behavior, including the availability of all associated outputs.
2. *Real-Time System*: A system in which logical correctness is based on both the correctness of the outputs and their timelines.
3. *Hard Real-Time System*: A system in which failure to meet even a single deadline may lead to complete or catastrophic system failure.
4. *Firm Real-Time System*: A system in which a few missed deadlines will not lead to total failure but missing more than a few may lead to complete or catastrophic system failure.
5. *Soft Real-Time System*: A system in which performance is degraded but not destroyed by failure to meet response-time constraints (Laplante & Ovaska, 2012).

These traditional measures of performance can be recommended as building blocks for next-generation IoT trust metrics. For example, taking a weighted average of response times across a set of actuation and event combinations can give a "response time" for an IoT system. Once "response time" is defined, then notions of deadline satisfaction and designation of hard, firm, or soft real-time can be assigned. Furthermore, repositories of performance data for various types of IoT systems, devices, and communication channels should be created for benchmarking purposes and eventual development of standards.

10.19 USABILITY

One of the larger concerns in IoT trust is usability – the extent to which a product can be used by specified users to achieve specified goals with effectiveness, efficiency, and satisfaction in a specified context of user – essentially, how "friendly" devices are to use and learn. This factor is an important consideration for most IT systems but may be more of a challenge with IoT, where the user interface may be tightly constrained by limited display size and functionality, or where a device can only be controlled via remote means. User interfaces for some device classes, such as smart home devices, are often limited to a small set of onboard features (e.g., LED status indicators and a few buttons) and a broader set of display and control parameters accessible remotely via a computer or mobile device. Some "smart" household items, such as light bulbs or faucets, may have no direct interface on the device and must be managed through a computer or smartphone connected wirelessly.

Such limited interfaces have significant implications for user trust. How do users know what action to take to produce a desired response, and how does the device issue a confirmation that will be understood? Devices with only a small display and one or two buttons often end up requiring complex user interactions that depend on sequences and timing of button presses or similar nonobvious actions. Consequently, many basic security functions can only be accomplished using a secondary device such as a smartphone. For example, if the IoT device has only two buttons, a password update will have to be done through the secondary device. As a result of this usability problem, users become even less likely to change default passwords, leaving the device open to attack. This is just one example of the interplay between usability and other trust factors. In the discussion below, we illustrate some of the complex interactions between usability engineering and factors such as performance, security, and synchronization.

Limited interfaces may to some extent be unavoidable with small devices but go against secure system principles harkening to Kerckhoffs' rules for crypto systems from the 19th century (Kerckhoffs, 1883) and later extended for IT systems (Saltzer & Schroeder, 1975). Among these is the principle that a secure system must be easy to use and not require users to remember complex steps. IoT systems run counter to this principle by their nature. Today, device makers are inventing user interfaces that often vary wildly from device to device and manufacturer to manufacturer, almost ensuring difficulty in remembering the right steps to follow for a given device.

One of the challenges of designing for IoT usability is the asynchronous operation imposed by device processing and battery limitations. Since devices may only be able to communicate periodically, with possibly minutes to hours between transmissions, conditions at a given time may be different than indicated by the last data received from a device. And since decision triggers may require readings from multiple devices, it is likely that decisions may be based on at least some currently invalid values, or actions may be delayed as the system waits for updated values. In the worst case, badly implemented IoT can "make the real world feel very broken" (Treseler, 2014), as when flipping a light switch results in nothing happening for some time as devices communicate.

In efforts to reduce usability problems, manufacturers have turned to AI to allow users to interact with their devices. One of the most popular uses of AI is for smart speakers, which allow users to simply talk to the device to control household appliances or order products. But as with much IT, building in one feature can have significant implications for others. Implementing the desirable voice-response capability requires engineering trade-offs that inherently impact the ability to build-in security. In one well-known example, a smart speaker responded to an accidental sequence of trigger words to record a conversation, then sent the recording to someone else (Soper, 2018). Consider the functions required for building such a system. The most obvious implication is of course that the device must be listening continuously to be able to respond without a user flipping an ON switch. And since small devices don't have the processing capacity and databases required for voice recognition, data must be sent to a larger processor in a cloud or similar service. Consequently, there is an always-active listening device with a connection to the Internet, clearly a security risk that will be challenging to defend against. Added to this is the need to prevent the system from misinterpreting user directions that it hears (consider how difficult it is sometimes to prevent misunderstandings between two people communicating). It is easy to see how building in usability can sometimes lead to real challenges in providing effective security.

One recommendation to improve usability for devices is to provide consistency among user interfaces. Standardized approaches will need to be developed, similar to what occurred for graphical user interfaces (GUIs) in the 1980s–1990s. Prior to the 1980s, computer interfaces were typically limited to keyboards and text displays with some basic graphical capabilities. Today's desktop GUIs have reasonably consistent "WIMP" (windows, icons, menus, and pointers) style interfaces that behave similarly across GUIs from different vendors, such as the desktop metaphor, double-clicking to open files, and drag-and-drop functions to manipulate objects. But this standardized interface was reached only after a decade or more of conflicting design and industry standards development. A similar process will be needed for IoT but will be longer and more difficult given the wide range of device types.

10.20 VISIBILITY AND DISCOVERABILITY

More than anything else, IoT represents the merger of information and communications technology with the physical world. This is an enormous change in the way that humans relate to technology, whose full implications will not be understood for many years. As with many aspects of technology, the change has been occurring gradually for some time but has now reached an exponential growth phase. However, by its nature, this merger of information technology with the physical world is not always obvious. Mark Weiser, who coined the term Ubiquitous Computing nearly 30 years ago, said that "The most profound technologies are those that disappear. They weave themselves into the fabric of everyday life until they are indistinguishable from it" (Weiser, 1991). Today this vision is coming true, as IoT devices proliferate into every aspect of daily life. According to one study, within 4 years, there will be more than 500 IoT devices in an average household (Gartner), so that they truly are beginning to disappear.

But is this disappearance uniformly a good thing? If a technology is invisible, then users will not be aware of its presence, or what it is doing. Trust issues related to this new technology world made news when reports suggested that smart televisions were "eavesdropping" on users (Tsukayama, 2015; Chung et. al., 2017). Voice-operated remote controls in smart televisions can only work if the televisions are always "listening", but the trust implications are obvious. To resolve trust concerns in cases like this, appliances need to be configurable for users to balance convenience with their personal security and privacy requirements, and device capabilities need to be visible with clear explanation of implications.

A different set of trust concerns is involved with technical aspects of device discovery in networks of 'things'. The traditional Internet was built almost entirely on the TCP/IP protocol suite, with HTML for web sites running on top of TCP/IP. Standardized communication port numbers and internationally agreed web domain names enabled consistent operation regardless of the computer or router manufacturer. Smartphones added the Bluetooth protocol for devices. This structure has not extended to IoT devices, because they generally do not have the processing power to support it. Instead, a proliferation of protocol families has developed by different companies and consortia, including BLE, ZigBee, Digital Enhanced Cordless Telecommunications Ultra Low Energy (DECT ULE), and a collection of proprietary technologies for Low Power Wide Area Networks (LPWAN). These many technologies result in a vast number of possible interactions among various versions of software and hardware from many different sources.

Most computer users are familiar with problems that arise when some business application or other software will not run because other software was changed on the system, and the two packages are no longer compatible. At least with PCs and mainframes, a person generally has a good idea of what is running on the systems.

With 500 IoT devices in a home, will the homeowner even know where the devices are located? How do devices make their presence known, with multiple protocols? It may not be clear from day to day what devices are on a network, or where they are, much less how they are interacting.

Device discovery is a complex problem for networks of things (Bello, Zeadally & Badra, 2017; Sunthonlap et al., 2018), but the general problem of discovery within networks has been studied for decades. There are generally two approaches:

Centralized: Nodes register with a central controller when they are brought into a network. The controller manages a database of currently available devices and periodically sends out heartbeat messages to ensure devices are available, dropping from the database any that don't respond.

Distributed: In this case, devices conduct a search for partner devices with the necessary features, by broadcasting to the local network. This approach avoids the need for a central controller, providing flexibility and scalability.

Scalability requirements for networks of hundreds of things often lead to implementing the distributed approach, but trust issues have enormous implications for device discovery in a large network. Without sophisticated cryptographically based authentication mechanisms, it becomes very difficult to ensure trusted operation in a network. For example, it has been shown that malware installed on a smartphone can open paths to other IoT devices, leaving the home network fully vulnerable to attack (Sivaraman et al., 2016). This is possible primarily because many IoT devices have little or no authentication, often due to the resource constraints described earlier.

Discoverability of IoT devices is thus a key problem for trust. Its dimensions include human factors, such as user's trust in behavior of devices such as the smart TV example, and technical issues of authentication among devices. Solutions will require adoption of some common protocols, which may take years for development of consensus standards, or emergence of de facto proprietary standards. In many cases, there will also be organizational challenges, since different kinds of devices may be installed by different departments. Organizations will need to know what devices are present, to manage security, or even just to avoid duplication of effort. This need can be addressed with audit tools that can identify and catalog devices on the network, reducing dependence on user cooperation but requiring trust in the audit tools.

10.21 SUMMARY

This publication has enumerated 17 technical trust concerns for any IoT system based on the primitives presented in NIST (2016). These systems have significant differences with traditional IT systems, such as much smaller size and limited performance, larger and more diverse networks, minimal or no user interface, lack of consistent access to reliable power and communications, and many others. These differences necessitate new approaches to planning and design. An essential aspect of developing these new systems is understanding the ways in which their characteristics can affect user trust and avoiding a "business as usual" approach that might be doomed to failure in the new world of IoT.

For each of the technical concerns, this publication introduced and defined the trust issues, pointed out how they differ for IoT as compared with traditional IT systems, gave examples of their effect in various IoT applications, and when appropriate, outlined solutions to dealing with the trust issues. Some of these recommendations apply not only to IoT systems but to other traditional IT systems as well. For some of the trust issues, IoT introduces complications that defy easy answers in the current level of development. These are noted as requiring research or industry consensus on solutions. This document thus offers an additional benefit of providing guidance towards a roadmap on needed standards efforts or research into how to better trust IoT systems.

10.22   ADDITIONAL SUPPORTING INFORMATION

10.22.1   INSURABILITY AND RISK MEASUREMENT

IoT trust issues truly come to the fore in assessing the impact of this new technology on insurability and risk management, because insurance requires that risk be measured and quantified. In this area, the emergence of IoT can have significant trade-offs – networks of 'things' can make it easier to estimate risk for the physical systems in which devices are embedded, but estimating risk for the device networks themselves may be much more difficult than for conventional IT systems.

Cars, homes, and factories with embedded sensors provide more data than ever, making it possible to estimate their risks more precisely, a huge benefit for insurers (Forbes, 2016). For example, auto insurance companies have begun offering lower rates for drivers who install tracking devices in their vehicles, to report where, how, and how fast they drive. Depending on a user's privacy expectations, there are obvious trust issues, and the legal aspects of employers installing such devices to monitor employee driving are just now being developed (Grossenbacher, 2018). Additionally, an often, neglected aspect of such devices is the possible trade-off between reducing risk by measuring the physical world, such as with driving, and potential increased risk from a complex network of things being introduced into a vehicle or other life-critical system. Already there have been claims that vehicle tracking devices have interfered with vehicle electronics, possibly leading to dangerous situations (Neilson, 2014). Examples include claims of losing headlights and taillights unexpectedly and complete shutdown of the vehicle (Horcher, 2014), as a result of unexpected interactions between the vehicle monitor and other components of the car's network of things.

In addition to estimating the risk, and thus insurability, of systems with embedded IoT devices, cybersecurity risks may become much harder to measure. Quantifying potential vulnerability even for conventional client–server systems, such as e-commerce, is not well understood, and reports of data loss are common. As a result, insurance against cybersecurity attacks is expensive – a $10 million policy can cost $200,000/year, because of the risk (*Wall Street Journal*, 2018). It will be much more difficult to measure risk for IoT networks of thousands of interacting devices than it is even for a corporate system made up of a few hundred servers and several thousand client nodes. IoT interactions are significantly more varied and more numerous than standard client–server architectures. Risk estimation for secure systems requires measurement of a *work factor*, the time, and resource cost of defeating a security measure. The same principle has been applied to vaults and safes long before the arrival of IT systems – the cost of defeating system security must be much higher than the value of the assets protected, so that attackers have no motivation to attempt to break in. The problem for networks of things is that there are few good measures of the work factor involved in breaking into these systems. They are not only new technology, but they have vast differences depending on where they are applied, and it is difficult to evaluate their defenses.

From a protection cost standpoint, IoT systems also have a huge negative trade-off – the typical processor and memory resource limitations of the devices make them easier to compromise, while at the same time, they may have data as sensitive as what's on a typical PC, or in extreme cases may present risks to life and health. Implantable medical devices can be much harder to secure than a home PC, but the risks are obviously much greater (Newman, 2017; Rushanan et al., 2014). Determining the work factor in breaking security of such devices and "body area networks" is an unsolved problem. A basic goal may be to ensure that life-critical IoT devices adhere to sound standards for secure development (Haigh & Landwehr, 2015), but estimating risk for such systems is likely to remain a challenge.

To complicate matters further, IoT systems often provide functions that may inspire *too much* trust from users. Drivers who placed unwarranted trust in vehicle autonomy have already been involved in fatal crashes, with suggestions that they were inattentive and believed the car could successfully avoid any obstacle (Siddiqui & Laris, 2018). Establishing the *right level of trust* for users will likely be a human factor challenge with IoT systems for many years to come.

No specific recommendations are made here. It is inevitable that insurers and systems engineers will eventually develop appropriate risk measures and mitigation strategies for IoT systems.

10.22.2 REGULATORY OVERSIGHT AND GOVERNANCE

Regulations have the power to significantly shape consumer interaction with technologies. Consider motor vehicles, whose safety is regulated by the National Highway Traffic Safety Administration (NHTSA, 2018). NHTSA enforces the Federal Motor Vehicle Safety Standards, which specify minimum safety compliance regulations for motor vehicles to meet; notable stipulations include requiring seatbelts in all vehicles, which can help reduce fatalities in the case of vehicular accidents. NHTSA likewise licenses vehicle manufacturers – helping regulate the supply of vehicles that consumers can buy – and also provides access to a safety rating system that consumers can consult. Multiple studies have shown the potential for regulations to continue to increase the safety of motor vehicles (e.g., Neely & Richardson, 2009).

Regulatory oversight and governance have been established in most domains for safety-critical systems. However, there is no parallel to the NHTSA for IoT systems:

1. There are no regulations on the security of IoT devices.
2. There is no oversight on the licensing of IoT device manufacturers.
3. There are no governing authorities evaluating the security of IoT devices.

These problems are compounded due to the economies behind IoT: the barrier to entry to constructing an IoT device is low, meaning that the market contains many different devices and models from many different manufacturers, with very few authoritative bodies attesting to the security of any of these devices. While these problems extend into the traditional computing market – i.e., laptops and personal computers – the market mechanics have since driven most products towards consolidated products and features, making it easier for consumers to evaluate and understand the security offered by the devices and manufacturers.

Nonetheless, while there is no central entity regulating the security of IoT devices, recent progress has been seen as regulatory participants consider how they want to approach this complex problem. As an example, the Internet of Things Cybersecurity Improvement Act (Weaver, 2017) was introduced in 2017 with the goal of setting standards for IoT devices specifically installed in government networks. The bill contains several important stipulations, including requiring devices to abandon fixed, default passwords and that devices must not have any known vulnerabilities. The act also relaxes several other acts that could be used to prosecute security researchers looking to test the safety of these devices.

The mandates of several agencies border with the IoT security space. A good example of this is the Federal Trade Commission (FTC). In January 2018, the VTech Electronics agreed to settle charges by the FTC that they violated not a security law, but rather US children's privacy law, collecting private information from children, not obtaining parental consent, and failing to take reasonable steps to secure the data (Federal Trade Commission, 2018). The key phrase is that last point: VTech's products were the Internet-connected toys – i.e., IoT devices – which collected personal information, and due to security risks in how these devices handled and managed data, the company was fined. This case shows that if IoT devices don't have reasonable security, a manufacturer may be held liable.

The US Consumer Product Safety Commission has called for more collaboration between lawyers and experts in the area (American Bar Association, 2017). Outside of the United States, the European Union Agency for Network and Information Security (ENISA) has published recommended security guidelines for IoT (ENISA, 2017). As more calls for security and recommendations occur, standardization and regulation may follow, increasing the security and safety of deployed IoT systems.

Regulations offer a serious means with which can help increase the security and safety of IoT systems, as evidenced by their successes in other industries such as vehicle manufacturing. While some improvements have been noticed as some agencies and organizations attempt to wield influence in IoT regulation, it has not been seen where any one central organization mandates rules regarding the use and development of IoT systems. Such an organization could have significant positive impact on the security and safety of IoT systems and consumers' lives.

10.22.3 Six Trustworthiness Elements in NIST SP 800-183

Six trustworthiness elements are listed in Section 10.3 of NIST SP 800-183. The verbatim text for those six is given here, and note that NoT stands for network of 'things':

[begin verbatim text]

To complete this model, we define six elements, namely, *environment*, *cost*, *geographic location*, *owner*, *Device_ID*, and *snapshot*, that although are not primitives, are key players in trusting NoTs. These elements play a major role in fostering the degree of trustworthiness[7] that a specific NoT can provide.

1. *Environment*: The universe that all primitives in a specific NoT operate in; this is essentially the *operational profile* of a NoT. The environment is particularly important to the sensor and aggregator primitives since it offers context to them. An analogy is the various weather profiles that an aircraft operates in or a particular factory setting that a NoT operates in. This will likely be difficult to correctly define.

2. *Cost*: The expenses, in terms of time and money, that a specific NoT incurs in terms of the nonmitigated reliability and security risks; additionally, the costs associated with each of the primitive components needed to build and operate a NoT. Cost is an estimation or prediction that can be measured or approximated. Cost drives the design decisions in building a NoT.

3. *Geographic Location*: Physical place where a sensor or *e*Utility operates in, e.g., using RFID to decide where a 'thing' actually resides. Note that the operating location may change over time. Note that a sensor's or *e*Utility's geographic location along with communication channel reliability and data security may affect the dataflow throughout a NoT's workflow in a timely manner. Geographic location determinations may sometimes not be possible. If not possible, the data should be suspect.

4. *Owner*: Person or organization that owns a particular sensor, communication channel, aggregator, decision trigger, or *e*Utility. There can be multiple owners for any of these five. Note that owners may have nefarious intentions that affect overall trust. Note further that owners may remain anonymous. Note that there is also a role for an **operator**; for simplicity, we roll up that role into the owner element.

5. *Device_ID*: A unique identifier for a particular sensor, communication channel, aggregator, decision trigger, or *e*Utility. Further, a Device_ID may be the only sensor data transmitted. This will typically originate from the manufacturer of the entity, but it could be modified or forged. This can be accomplished using RFID[8] for physical primitives.

6. *Snapshot*: An instant in time. Basic properties, assumptions, and general statements about snapshot include:
   a. Because a NoT is a distributed system, different events, data transfers, and computations occur at different snapshots.

---

[7] *Trustworthiness* includes attributes such as security, privacy, reliability, safety, availability, and performance, to name a few.

[8] RFID readers that work on the same protocol as the inlay may be distributed at key points throughout a NoT. Readers activate the tag causing it to broadcast radio waves within bandwidths reserved for RFID usage by individual governments internationally. These radio waves transmit identifiers or codes that reference unique information associated with the item to which the RFID inlay is attached, and in this case, the item would be a primitive.

b. Snapshots may be aligned to a clock synchronized within their own net-work (NIST, Weiss et al., 2015). A global clock may be too burdensome for sensor networks that operate in the wild. Others, however, argue in favor of a global clock (Li, 2004). This publication does not endorse either scheme at the time of this writing.

c. Data, without some "agreed upon" time stamping mechanism, is of limited or reduced value.

d. NoTs may affect business performance – sensing, communicating, and computing can speed up or slow down a NoT's workflow and therefore affect the "perceived" performance of the environment it operates in or controls.

e. Snapshots maybe tampered with, making it unclear when events actually occurred, not by changing time (which is not possible) but by changing the recorded time at which an event in the workflow is generated, or computation is performed, e.g., sticking in a **delay()** function call.

f. Malicious latency to induce delays is possible and will affect when decision triggers are able to execute.

g. Reliability and performance of a NoT may be highly based on (e) and (f).

[end verbatim text]

NISTIR 8222 (DRAFT) has taken Section 10.3 from NIST SP 800-183 and expanded into a richer discussion as to why trusting IoT products and services is difficult. It has derived 17 new technical trust concerns from the six elements in NIST SP 800-183. For example, the snapshot element briefly mentioned in NIST SP 800-183 is discussed in detail in Section 10.7 concerning a lack of precise timestamps.

ACRONYM GLOSSARY

AI	Artificial Intelligence
BBC	British Broadcasting Corporation
BLE	Bluetooth Low Energy
COTS	Commercial Off-the-Shelf
DECT ULE	Digital Enhanced Cordless Telecommunications Ultra Low Energy
ENISA	European Union Agency for Network and Information Security
FTC	Federal Trade Commission
GPS	Global Positioning System
HTML	Hypertext Markup Language
HTTPS	Hypertext Transfers Protocol Secure
IETF	Internet Engineering Task Force
IIOT	Industrial Internet of Things
IoT	Internet of Things
IT	Information Technology
LPWAN	Low Power Wide Area Network
MUD	Manufacturer Usage Description
NHTSA	National Highway Traffic Safety Administration
NIST	National Institute of Standards and Technology

NoT	Network of Things
PC	Personal Computer
RFID	Radio Frequency Identification
SLOC	Source Lines of Code
TCP/IP	Transmission Control Protocol/Internet Protocol

REFERENCES

American Bar Association, "Consumer Product Safety Administration seeks collaboration in managing internet of things," May 2017. https://www.americanbar.org/news/abanews/aba-news-archives/2017/05/consumer_productsaf.html. Accessed July 21, 2018.

Ashton, K., "That 'Internet of Things' thing," June 22, 2009. http://www.rfidjournal.com/articles/view?4986, retrieved 5/9/2017.

BBC, "Peter Day's world of business". BBC World Service, 2016. http://downloads.bbc.co.uk/podcasts/radio/worldbiz/worldbiz_20150319-0730a.mp3, retrieved 8/4/2016.

Bello, O., Zeadally, S., Badra, M., "Network layer inter-operation of Device-to-Device communication technologies in Internet of Things (IoT)," *Ad Hoc Networks*, vol. 57, 2017, pp. 52–62.

Chung, H., Iorga, M., Voas, J., "Alexa, can i trust you?," *IEEE Computer*, vol. 50, no. 7, 2017, pp. 100–104.

Columbus, L., "Internet of Things market to reach $267B by 2020", January 29, 2017. https://www.forbes.com/sites/louiscolumbus/2017/01/29/internet-of-things-market-to-reach-267b-by-2020/#f2ddc5609bd6, retrieved 1/5/2018.200

Coraggio, G., "The Internet of Things and its legal dilemmas," VC Experts Blog, December 15, 2016. https://blog.vcexperts.com/2016/12/15/the-internet-of-things-and-its-legal-dilemmas.

Data IQ News, "Big data to turn 'mega' as capacity will hot 44 zettabytes by 2020," October 2014. http://www.dataiq.co.uk/news/20140410/big-data-turn-mega-capacity-will-hit-44-zettabytes-2020.

Dhadyalla, G., Kumari, N., Snell, T., "Combinatorial testing for an automotive hybrid electric vehicle control system: a case study," *Proceedings of the IEEE 7th International Conference on Software Testing, Verification and Validation Workshops (ICSTW 14)*, Cleveland, OH, 2014, pp. 51–57.

European Union Agency for Network and Information Security, "Baseline security recommendations for IoT," November 2017. https://www.enisa.europa.eu/publications/baseline-security-recommendations-for-iot. Accessed July 21, 2018.

Federal Trade Commission, "Electronic toy maker VTech settles FTC allegations that it violated children's privacy law and the FTC Act," January 2018. https://www.ftc.gov/news-events/press-releases/2018/01/electronic-toy-maker-vtech-settles-ftc-allegations-it-violated. Accessed July 21, 2018.

Forbes, 2016. https://www.forbes.com/sites/robertreiss/2016/02/01/5-ways-the-iot-will-transform-the-insurance-industry/.

Gartner, 2015. ttp://www.gartner.com/newsroom/id/2839717.

Grossenbacher, K., "The legality of tracking employees by GPS," 2018. https://www.laborandemploymentlawcounsel.com/2016/01/the-legality-of-tracking-employees-by-gps/.

Haigh, T, Landwehr, C., "Building code for medical device software security," *IEEE Cybersecurity*, 2015. https://www.ecirtam.net/autoblogs/autoblogs/lamaredugoffrblog_6aa4265372739b936776738439d4ddb430f5fa2e/media/1fe94efd.building-code-for-medica-device-software-security.pdf

Horcher, G., "Concerns with insurance devices that monitor for safe-driver discounts," *Atlanta Journal Constitution*, 2014.

Kerckhoffs, A., "La cryptographie militaire," *Journal des sciences militaires*, vol. IX, 1883, pp. 5–83, January, pp. 161–191, February. https://en.wikipedia.org/wiki/Kerckhoffs%27s_principle.

Kolias, C., Kambourakis, G., Stavrou, A., Voas, J., "DDoS in the IoT: Mirai and other botnets,". *Computer*, vol. 50, no. 7, 2017, pp. 80–84.

Laplante, P. A., Ovaska, S. J., *Real-Time Systems Design and Analysis*. 4th Edn., John Wiley & Sons/IEEE Press, Piscataway, NJ, 2012.

Lear, E., Droms, R., Romascanu, D., "Manufacturer usage description specification," IETF draft, 2017.

Li, Q. and Rus, D., "Global clock synchronization in sensor networks," Twenty-third Annual Joint Conference of the IEEE Computer and Communications Societies (INFOCOM 2004), Hong Kong, March 7–11, 2004, pp. 564–574. http://dx.doi.org/10.1109/INFCOM.2004.1354528.

Miller, K. W., Voas, J., "Software certification services: encouraging trust and reasonable expectations", *IEEE IT Professional*, vol. 8, no. 5, 2006, pp. 39–44.

Miller, K. W., Voas, J., Laplante, P. A., "In trust we trust," *IEEE Computer*, vol. 43, 2010, pp. 91–93.

Moran, B., Meriac, M., Tschofenig, H., "A firmware update architecture for Internet of Things devices," 2017. https://tools.ietf.org/id/draft-moran-suit-architecture-00.html.

Musa, J., Iannino, A., Okumoto, K., *Software Reliability: Measurement, Prediction, Application*. McGraw-Hill, Singapore, 1987.

National Highway Traffic Safety Administration, 2018. https://www.nhtsa.gov/. Accessed July 20, 2018.

Neeley, G. W., Richardson Jr, L. E., "The effect of state regulations on truck-crash fatalities," *American Journal of Public Health*, vol. 99, no. 3, 2009, pp. 408–415.

Neilson, "Insurance tracking device blamed for car damage," 2014. https://www.program-business.com/News/Insurance-Tracking-Device-Blamed-for-Car-Damage.

Neumann, P. G., "Risks forum," June 11, 2018.

Newman, L. H., "Medical devices are the next security nightmare," *WIRED*, Np, 2017.

Patil, A. H., Goveas, N., Rangarajan, K., "Test suite design methodology using combinatorial approach for Internet of Things Operating Systems," *Journal of Software Engineering and Applications*, vol. 8, no. 7, 2015, p. 303.

Rushanan, M., Rubin, A. D., Kune, D. F., Swanson, C. M., "SoK: Security and privacy in implantable medical devices and body area networks," *2014 IEEE Symposium on Security and Privacy (SP)*, Berkeley, CA, pp. 524–539, 2014.

Salman, T., "Internet of Things protocols," November 30, 2015. https://www.cse.wustl.edu/~jain/cse570-15/ftp/iot_prot/.

Saltzer, J. H., Schroeder, M. D., "The protection of information in computer systems," *Proceedings of the IEEE*, vol. 63, no. 9, 1975, pp. 1278–1308.

Siddiqui, F., Laris, M., "Self-driving Uber vehicle strikes and kills pedestrian," Washington Post, March 19, 2018. https://www.washingtonpost.com/news/dr-gridlock/wp/2018/03/19/uber-halts-autonomous-vehicle-testing-after-a-pedestrian-is-struck/.

Sivaraman, V., Chan, D., Earl, D., Boreli, R., "Smart-phones attacking smart-homes," *Proceedings of the 9th ACM Conference on Security & Privacy in Wireless and Mobile Networks*, Darmstadt, Germany, pp. 195–200, July 2016.

Stavrou, A., Voas, J., "Verified time," *IEEE Computer*, vol. 50, no. 3, 2017, pp. 78–82.

Soper, S., "This is how Alexa can record private conversations," 2018. https://www.bloomberg.com/news/articles/2018-05-24/amazon-s-alexa-eavesdropped-and-shared-the-conversation-report.

Sunthonlap, J., Nguyen, P., Wang, H., Pourhomanyoun, M., Zhu, Y., Ye, Z., "SAND: a social-aware and distributed scheme for device discovery in the Internet of Things," *2018 International Conference on Computing, Networking and Communications (ICNC)*, Maui, HI, pp. 38–42, March 2018.

Treseler, M., "How is UX for IoT different?," 2014. http://radar.oreilly.com/2014/11/how-is-ux-for-iot-different.html.

Tsukayama, H., "Samsung: our televisions aren't secretly eavesdropping on you," 2015. https://www.washingtonpost.com/news/the-switch/wp/2015/02/10/samsung-our-televisions-arent-secretly-eavesdropping-on-you/?noredirect=on&utm_term=.5322af153a88.

Voas, J., "Error propagation analysis for COTS systems," *IEEE Computing and Control Engineering Journal*, vol. 8, no. 6, 1997, pp. 269–272.

Voas, J., "Certifying off-the-shelf software components," *IEEE Computer*, vol. 31, no. 6, 1998a, pp. 53–59. (Translated into Japanese and reprinted in Nikkei Computer magazine)

Voas, J., "The software quality certification triangle," *Crosstalk*, vol. 11, no. 11, 1998b, pp. 12–14.

Voas, J., "Certifying software for high assurance environments," *IEEE Software*, vol. 16, no. 4, 1999, pp. 48–54.

Voas, J., "Toward a usage-based software certification process," *IEEE Computer*, vol. 33, no. 8, 2000, pp. 32–37.

Voas, J., "Software's secret sauce: the 'ilities'," *IEEE Software*, vol. 21, no. 6, 2004, pp. 14–15.

Voas, J., "Networks of 'Things'," *NIST Special Publication*, vol. 800, 2016, p. 183.

Voas, J., Hurlburt, G., "Third party software's trust quagmire," *IEEE Computer*, vol. 48, 2015, pp. 80–87.

Voas, J., Laplante, P. A., "The IoT blame game," *IEEE Computer*, vol. 50, no. 6, 2017, pp. 69–73.

Voas, J., Payne, J., "Dependability certification of software components," *Journal of Systems and Software*, vol. 52, 2000, pp. 165–172.

Voas, J., Charron, F., Miller, K., "Tolerant software interfaces: can COTS-based systems be trusted without them?," *Proceedings of the 15th International Conference on Computer Safety, Reliability and Security (SAFECOMP'96)*, Springer-Verlag, Vienna, Austria, pp. 126–135, October 1996.

Voas, J., Kuhn, R., Laplante, P. A., "Testing IoT systems," *2018 IEEE Symposium on Service-Oriented System Engineering (SOSE)*, Bamberg, Germany, pp. 48–52, March 2018a.

Voas, J., Kuhn, R., Laplante, P. A., "IoT metrology", *IEEE Annals of the History of Computing*, vol. 20, no. 3, 2018b, pp. 6–10.

Wall Street Journal, 2018. https://www.wsj.com/articles/ransom-demands-and-frozen-computers-hackers-hit-towns-across-the-u-s-1529838001.

Weaver, N., "The Internet of Things Cybersecurity Improvement Act: a good start on IoT security," August 2017. https://www.lawfareblog.com/internet-things-cybersecurity-improvement-act-good-start-iot-security. Accessed July 21, 2018.

Weiser, M., "The computer for the 21st century," *Scientific American*, vol. 265, no. 3, 1991, pp. 94–105.

Weiss, M., Eidson, J. Barry, C., Broman, D., Goldin, L., Iannucci, B., Lee, E. A., and Stanton, K., Time-Aware Applications, Computers, and Communication Systems (TAACCS), NIST Technical Note (TN) 1867, National Institute of Standards and Technology, Gaithersburg, February 2015, 26pp. http://dx.doi.org/10.6028/NIST.TN.1867

Yang, J., Zhang, H., Fu, J., "A fuzzing framework based on symbolic execution and combinatorial testing," *Green Computing and Communications (GreenCom), 2013 IEEE and Internet of Things (iThings/CPSCom), IEEE International Conference on Cyber, Physical and Social Computing*, Beijing, China, pp. 2076–2080, August 2013.

11 Blockchain Technologies and IoT[1]

11.1 INTRODUCTION

According to Hileman and Rauchs (2017) "A blockchain is a special data structure that is composed of transactions, batched into blocks, that are cryptographically linked to each other to form a sequential, tamper-evident chain event that determines the ordering of transactions in the system. In this context, a transaction represents any change or modification to the database". More broadly, a blockchain is a peer-to-peer (P2P) distributed network of independent participants that generally broadcast all data to each other, each of whom may have different motivations and objectives. They may not necessarily trust one another but reach a consensus (a consistent agreement about changes to the state of the shared database) on a linear history of operations of that shared database. A high-level workflow of blockchain is presented in Figure 11.1.

The key advantages of blockchain (compared to existing distributed systems and database technologies) are the use of a specialized data structure that bundles transactions into blocks and the broadcast of all data to all participants. Some of the main components of a blockchain are cryptography, P2P networks, consensus mechanisms, the ledger, validity rules, and access or permission types.

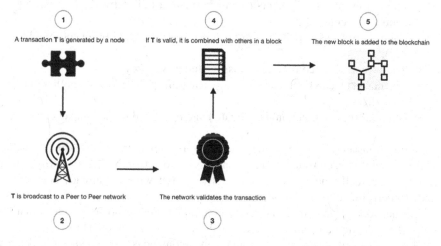

FIGURE 11.1 Blockchain technology.

---

[1] This chapter was contributed by Giuseppe Destefanis, Lodovica Marchesi, Michele Marchesi, Marco Ortu, Roberto Tonelli.

DOI: 10.1201/9781003027799-11

There are general permission type distinctions for current blockchain architectures:

- Permissionless, public, or open refers to blockchains where access is not restricted to a specific set of vetted participants. In these types of blockchain, participants do not know and trust each other;
- "Permissioned", "private", or "closed" refers to blockchains where access is restricted to a specific set of vetted participants (Hileman & Rauchs, 2017). These blockchains operate in an environment where participants are already known, and there is a level of trust among them; this removes the need for a native token to incentivize good behavior. Participants are held liable through off-chain legal contracts and agreements and are incentivized to behave honestly via the threat of legal prosecution in the case of misbehavior (Hileman & Rauchs, 2017).
- "Consortium" or "federated" refers to a blockchain where the architecture could be private or hybrid (public and private). This type of blockchain uses features such as permission restriction, multiple controlling authorities; they allow easy, yet controlled information sharing between various stakeholders and more.

11.2   THE IDEA OF CRYPTOCURRENCIES

The first application on a blockchain has been Bitcoin, the most famous cryptocurrency.

The basic ideas on which Bitcoin (and the other cryptocurrencies) is based are the following:

- The coin is a registration in a public register (the blockchain), shared on the Internet by thousands of nodes which create a P2P network, without any centralized controller.
- Each record has a public address (derived from a public cryptographic key).
- The owner of the coin owns the associated private key.
- The currency can be transferred only using this private key to another public address.
- This transfer (transaction) is validated and registered in the blockchain.

Verifying transactions and creating blocks has a computational cost which must be remunerated. For each transaction, it is necessary to verify that (1) the entity which activates it is really the owner, (2) the relative currency has not already been spent, and (3) there are no other errors.

The network is democratic and anonymous, and the nodes do not give information about their identity; anyone able to access the network can validate blocks.

However, an attacker could create many anonymous (thousands or millions) seemingly independent nodes and take control of the network. This attack is called "Sibling Attack", and a fundamental constraint is that the system must be able to work even in the presence of attacks and fraud attempts. Trusting all the participants in the network is not a requirement for the system to work.

The Sibling Attack problem is solved by making block validation computationally expensive. This concept is called "Proof of Work" (PoW). To validate the transactions, and therefore the blocks, it is necessary to have high-speed and therefore energy-consuming hardware. The network self-adjusts to the available computing power so that the time required for validating transactions is approximately constant. In this scenario, a Sibling Attack would be costly and therefore impractical.

Validation of transactions and blocks through PoW is also used for creating currency. The first node which validates the current block, stores it in the blockchain and earns a fixed amount of cryptocurrency. All nodes concurring for solving the computational challenge accept the validated block, stop solving it, and move on to the next block. In this way, two problems are solved:

- The remuneration of mining;
- The gradual introduction of new currency in the system.

Existing cryptocurrencies typically have a limited total amount of currency, or one that grows by a fixed amount each year, so that the growth percentage decreases over time. Bitcoin is an example of the first approach: the reward for validating a block is halved every 4 years, until it reaches zero in 2040. Ethereum is an example of the second approach.

11.3   BITCOIN

Bitcoin is the cryptocurrency par excellence, the mother of all other cryptocurrencies. It was introduced in 2009 by Satoshi Nakamoto (Nakamoto & Bitcoin, 2008), the pseudonym of a still unknown person or group. Bitcoin is used with three different meanings:

- The architecture/platform of the Bitcoin network;
- The protocols to make this platform work;
- The "currency" traded on the platform (the acronym is BTC).

One Bitcoin is composed of 100 million Satoshi (the current minimum transfer unit of the cryptocurrency).

11.3.1   THE BITCOIN ADDRESS

A Bitcoin address is a number associated with a private–public key pair, and it is the public part of all Bitcoin transfers.

An address is generated from the public key so that:

- An address corresponds to a single public key;
- Given the address, it is not possible to know the public key; this is because the explicit knowledge of this key could in the future make the system more vulnerable;
- An address contains checks in order to recognize any typing errors (as with the tax code).

When transferring the Bitcoins associated with one address to another, the public key of the first must be revealed.

The procedure for generating a new Bitcoin address is as follows:

1. A random number **d** (private key) of 256 bits is generated;
2. The public key **B** is generated starting from **d** and from the standard elliptic curve secp256k1;
3. Starting from **B**, the Bitcoin address **A** is generated with a procedure;
4. A Bitcoin address has about 195 bits, and it is represented with 30–34 alphanumeric characters (BASE58 representation) always starting with "1".

Given an address, only the owner of the key **d** can generate **B** from it, and from **B** generate **A**.

The steps to generate a Bitcoin address from the public key from **B** are the following:

1. Calculation of the RIPEMD160 hash of the SHA256 hash of B:
   hash: = RIPEMD160 (SHA256 (B))
2. Add a byte (address version) in the hash (head):
   ha: = b.hash with b equal to:
   0x05 (00000101) for the main net (Bitcoin)
   0xc4 (11000100) for the net test
3. Calculation of the SHA256 hash of the SHA256 hash of **ha**:
   hh: = SHA256 (SHA256 (ha))
4. Extract the first 4 bytes of **hh**, to be used as checksum (cs):
   cs: = First4Bytes (hh)
5. Concatenation of **b**, **hash** and **cs** $(8 + 160 + 32 = 200$ bits):
   a: = BASE58 (b.hash.cs)
   A BASE58 string is generated by encoding a number (a sequence of bits) in base 58, using the following 58 alphanumeric characters as digits from 0 to 57:
   123456789ABCDEFGHJKLMNPQRSTUVWXYZabcdefghijkmnop qrstuvwxyz
   Without the "ambiguous" characters: 0, O, I, 1

Examples of Bitcoin addresses:

1NdNaNG1K9eDW5jfMDmT1biyih5micPsGE
1yQ3KkhfhQCMABZkZxTdjAC2nGFUxoDcR

The procedure of calculating A starting from B, with four successive hashing is used to:

- Prevent key B from being made public (because it could be attacked with quantum computers in the future);
- Insert a "checksum" that verifies whether the address has been entered incorrectly: in this case, the software does not make payments to this address;

When the Bitcoins associated with an address are spent, the public key B is made visible in the blockchain, and thus, it becomes more vulnerable.

11.3.2 The Bitcoin Blockchain

The blockchain is composed of an ordered sequence of blocks. These blocks contain the validated transactions, which in turn contain the Bitcoin transfers from one address to another and the payment of the block validation reward to the address of the miner who validated it.

A block is composed of a header and all its transactions. A block header contains the following data:

- Date and time
- Number of transactions contained in the block
- Nonce (a 32-bit integer)
- Block hash
- Hash of the previous block
- Merkle tree hash containing transactions

It follows a list of all the transactions in the block, contained in a Merkle tree (an efficient data structure to verify by hash that the transactions have not been altered).

The blockchain is a sequence of blocks, each with a hash to ensure its inalterability. Each block also incorporates the hash of the previous block, thus "hooking" the blocks together and creating the "chain". If a block of the blockchain were altered, the hash would also be altered, and therefore, to preserve the integrity of the blockchain, it would also be necessary to modify all subsequent blocks. Since the computation of the hash of a block is computationally costly, such alteration is practically impossible.

The main features of a blockchain technology are as follows:

Transparency: all the transactions (the whole history) are stored forever and cannot be altered. For each transaction, it is possible to know the amount in Bitcoins, the addresses from which they are withdrawn, and to which the transfer goes. In this way, a complete tracking of Bitcoin transfer flows from one address to another is possible.

Anonymity: in the Bitcoin system, the holders of the funds are identified only by an anonymous address, to which a private key is associated. Thus, while the flow of Bitcoin from one address to another is completely transparent, the holders of the funds are only identified by an anonymous address.

Double Spending Protection: only the first verified transaction is accepted, all the others are rejected.

Nonrepudiation: once a transaction has been sent and accepted, it cannot be canceled for any reason.

Security: the transaction can only be activated by knowing the private key relating to the withdrawal address of the funds. If this key is lost, the related funds are lost forever.

Smart Contract: transactions can be activated by several parties involved (with their respective private keys), and they can carry out even complex preprogrammed calculations.

"Notarial" Storage: it is possible to use special transactions to store information on the blockchain. This information is used to certify the existence and integrity of a document, or set of documents, at a certain date.

11.3.3 THE BITCOIN TRANSACTION

The purpose of a transaction is to record the transfer of Bitcoin from one or more addresses to one or more other addresses in the blockchain.

Each transaction is uniquely identified by a hash, called its "id". A transaction has one or more inputs, and one or more outputs, although there are exceptions to this rule.

The inputs are the outputs of previous transactions not spent yet (UTXO, Unspent Transaction Output). Each input contains an address, the hash (id) of the transaction that transferred the Bitcoins to the address, and the amount transferred in that transaction (in Satoshi). All this information is verifiable because it is already present in the blockchain.

In output, there are one or more addresses, with the amount (in Satoshi) to be transferred for each address.

Typically, the sum of all Satoshis in input to the transaction is greater than the sum of those in output. The difference between input and output goes to the miner who validates the block as a "commission" (fee), and which is therefore added to the fixed fee in Bitcoin for validation. If the commission for the miners is zero, or too low, the transaction, while valid, may not be accepted or its validation may be delayed.

Transactions, once written on the blockchain, are irrevocable: the Bitcoin protocol does not provide any possibility for canceling a verified and recorded transaction. For a transaction to be valid, the following are necessary:

- All its inputs must be "signed" by the private keys associated with the addresses.
- None of the inputs must have already appeared as inputs in another transaction ("double spending" is not possible).
- The addresses of the outputs must be valid.
- The sum of the input amounts must be greater than or equal to that of the output amounts.

The inputs of a transaction are UTXO of a previous transaction. These inputs must all be spent within the transaction.

For transferring only part of the input to the output, it is necessary to add an output that transfers the "remaining amount" to an address controlled by whoever carries out the transaction itself. This address can be the one (or one of those) used for the input, but usually, a new address is generated for security and privacy reasons.

In each block, the first transaction contains a "dummy" input, called "coinbase". The first transaction contains an output, the address of the miner who validated the block, and is therefore a transaction entered by the miner itself.

The coinbase transfers a fixed amount of Bitcoin to the output as a "reward" for the validation, plus the commission fees coming from the block transactions. This fixed amount, initially 50 Bitcoins, halves every 4 years.

The Bitcoin protocol provides the possibility of more complex transactions than simply transferring currency to be executed. These transactions allow the execution of scripts whose success is linked to the validation of the entire transaction.

Among these, the simplest transaction is the "OP_RETURN" type transaction. It has no output and sends all the Bitcoin in input to the miner. It is used to record information (equal to 80 bytes) within the transaction itself.

The most obvious form of attack to the Bitcoin system is "Double Spending". It is an attempt to spend two or more times the output of a transaction.

The protocol protects itself from this attack by considering valid only the first transaction that spends the output and that is recorded in the blockchain. In doing so, the problem is solved in a decentralized way, without the need for a central authority to decide which transaction is the valid one.

The purpose of block validation is to:

- Validate transactions sent to the network
- Prevent fraudulent behavior

Validation is a difficult problem (on purpose), so that no one can validate a block in a short time (proof of work).

The mechanism works as follows:

1. The users create and send transactions to the network;
2. The miners verify the transactions, and if the fees are sufficient, insert them in their block to be validated;
3. The first miner who validates the block communicates it to the other miners, who insert this block in their blockchain and start validating the next block.

Transactions sent to the Bitcoin network for validation are called "unconfirmed transactions". If a transaction is placed in a block that is validated, it becomes "confirmed".

The nodes maintain a list of unconfirmed transactions, called unconfirmed transactions' memory pool or mempool.

To be added to the mempool, a transaction must still be validated and must not spend inputs already spent by a transaction already contained in the mempool. Transactions validated by orphan blocks, but not in the correct blockchain, are put back in the mempool. If a transaction grants a fee that miners consider too low, it remains in the mempool until it is withdrawn or until some miners put it in a block which is then validated.

Occasionally, two miners can validate a block almost simultaneously, communicating it to the network. In this case, some of the miners accept the first block and some accept the second. This phenomenon is called "fork" of the blockchain.

One part of the miners accepts the first block, the other part accepts the second block. However, as time passes, all the miners choose the longer branch of the blockchain, abandoning the shortest and invalidating the "orphan" blocks of that branch. Note that in case of double validation, the two competing blocks will contain almost the same transactions. It is probable that, in the two competing branches, the two subsequent blocks are not again validated almost simultaneously. In this case, one of the two branches becomes longer than a block, and the miners choose the chain containing it. Otherwise, the two branches will grow together, until one becomes longer than the other. The blockchain, however, also includes orphan blocks, which are however ignored in the validation of transactions.

11.4   SMART CONTRACTS

One of the fundamental properties of the blockchain is that it can be the basis for Smart Contracts. Smart Contracts are automated contracts, where participants prove their identity and agreement approval with their private key. Many of the major cryptocurrencies have been designed to be able to carry out Smart Contracts, albeit with different levels of sophistication.

Although the Smart Contract concept is old, the name and definition were proposed by Nick Szabo, an American law graduate cryptography expert, in 1994. According to his definition: "A Smart Contract is a computerized protocol for executing transactions, which executes the terms of a contract. The general objectives of a Smart Contract are to satisfy the usual contractual conditions (such as payment terms, rights, confidentiality, and even enforcement), to minimize both intentional and accidental exceptions, and to minimize the need for trusted intermediaries. Other economic objectives include the reduction of costs due to fraud, arbitration and enforcement costs, and other transaction costs".

A Smart Contract is basically a program, which runs on a secure trusted medium. It takes digital signatures of participants, and other information, from secure sources as input. In output, it transfers cryptocurrency amounts, can activate other contracts, record information, or connect to external systems. Smart Contracts also often have a user interface that emulates the jargon and logic of contract clauses.

Since the execution of a software program is deterministic and immutable, with the same input and program status, the code can be considered a contract. Once the contractual clauses are correctly inserted in the code of a Smart Contract, and this is accepted by the contractors, the effects are no longer linked to their will or to the action of intermediaries. Obviously, precise guarantees are needed:

- The code must have no errors, must be executed correctly, and must not be modified;
- The inputs to the code must come from secure and identified sources;
- The code outputs must achieve the desired effects.

In other words, there must be a mechanism that guarantees the trust that the contractors place in the Smart Contract. At this point, the blockchain comes into play: it provides all the previously mentioned guarantees, without the need for a central authority.

Since the first versions of the implementation of the Bitcoin blockchain, it was equipped with a real programming language, albeit at the assembler level. The Bitcoin transaction, which involves the transfer of cryptocurrency from one or more inputs to one or more outputs, is in fact carried out by executing code written in this language, which takes care to verify the authenticity of the private keys in possession of who carries out the transaction. However, the language of Bitcoin is limited, does not involve decision points or cycles, and is therefore not Turing complete.

The introduction of Ethereum in 2015 filled this gap, introducing a blockchain and a low-level language, along with various high-level and actually usable languages (among them we mention Solidity and Serpent), to code and execute Smart Contracts (Marchesi et al., 2020).

The best-known real implementation of a Smart Contract was the DAO. It was an autonomous distributed venture capital project funding organization, launched in May 2015 with US$150 million raised in the crowdfunding campaign. In June 2016, it was attacked and emptied of a third of the funds, corresponding to approximately US$50 million in ether cryptocurrency. The cause of this problem was an error in the writing of the Smart Contract, which left open a "door" capable of recursively emptying all the collected capital.

The debate that followed centered on the one hand that the feature of the code that enabled the attack clearly did not respect the will of the programmers and was a mistake, and on the other hand, the observation that, if the code is the contract, freely examined, and accepted by the parties, then this feature was also part of the contract, and the attack could not be considered a breach of contract. Since the very nature of Smart Contracts on blockchain does not provide for any authority with the power to intervene, it was the community of Ethereum developers and miners who decided by majority for the first hypothesis, up to the fork. As already mentioned, there were those who did not accept the fork and continued with the old blockchain and the old rules, resulting in the generation of Ethereum Classic.

Despite the DAO's epic failure, the topic of Smart Contracts is more relevant than ever. There are many initiatives, based on Ethereum, on extensions to the Bitcoin blockchain, on Ripple, or on ongoing implementations of new private blockchains, aimed at creating Smart Contract systems. The most active sector is certainly banking and finance (Dashkevich, Counsell & Destefanis, 2020), but also for other sectors, such as the Public Administration and the Internet of Things, we are actively working to find and experiment with Smart Contract applications.

Possible application sectors are electoral systems (Casaleggio et al., 2021), domain name registration, financial markets (Dashkevich, Counsell & Destefanis, 2020), healthcare (Kassab et al., 2019), crowdfunding platforms (Fenu et al., 2018), intellectual property, supply chain (Marchesi, Mannaro & Porcu, 2021).

11.5 IoT AND SMART CONTRACTS: A FOOD TRACEABILITY SYSTEM

Food fraud has become a global issue for producers, consumers, governments, and other involved actors. Although this is not a new problem, unfortunately, it is rising in the last few years and threatening both public health and the economy. According to Spink and Moyer (2011), "Food fraud is a collective term used to encompass the deliberate and intentional substitution, addition, tampering, or misrepresentation of food, food ingredients, or food packaging; or false or misleading statements made about a product, for economic gain".

Market globalization reduces food traceability facilitating food fraud. Consequently, food safety can be increased through higher traceability. To identify and address food frauds, it is essential to trace and authenticate the food supply chain to understand provenance and ensure food quality in compliance with international standards and national legislation.

Food traceability can be defined as "the ability to follow the movement of a feed or food through specified stage(s) of production, processing and distribution", as defined by the International Organization for Standardization in ISO 22005:2007 – a specific standard for traceability in the food and feed chain. Additionally, a traceability system is based on product labeling and contains both quantitative and qualitative information about the final product and its provenance.

In this context, a distributed ledger technology such as blockchain provides a complete and immutable audit trail of transaction data for all stages of the food supply chain, allowing for transparency and verifiable and immutable records in the form of digital certificates.

The immutability of the data enables the technology to be considered for regulated industries such as agri-food.

A decentralized application, or DApp (Marchesi, Marchesi & Tonelli, 2020), can be used for enabling food traceability. A *decentralized application* is a computer application that runs on a distributed system on a network of nodes, with no node acting as supervisor. In the blockchain P2P network, a DApp is stored and executed on the blockchain to be decentralized, transparent, deterministic, and redundant. A DApp is developed by writing Smart Contracts and may have a user interface that allows users and devices to interact with it.

Based on various definitions of traceability systems, the main purposes of a food traceability system are the following:

- To document transparently and irreversibly any event relevant to production;
- To allow authorities, laboratories, and certified experts to asseverate the production, giving proof of their identity and their certifications;
- To integrate manual registrations and automatic registrations, made by Internet of Things (IoT) devices that are increasingly widespread;
- To keep track of the quantities produced, so that these cannot be increased by introducing products of noncertified origin;
- To give evidence of all stages of production to the authorities responsible for verifying the specifications;
- To allow retailers and end consumers to learn about the history of the products purchased, from the field to the purchased product, using an app.

Traditional approaches for food traceability systems are usually inefficient. *Blockchain* is a promising technology that is tamper-proof and decentralized, and self-executing and self-verifying Smart Contracts can conduct transactions between mutually untrusted parties.

A traceability system based on blockchain would guarantee:

- Data integrity and provenance of documents and records on blockchain;
- Immutability and transparency of data recorded on the blockchain;
- Respect for the quantities of the products involved (grapes, wine, bottles) based on the annual production of the land and the yield in the various stages of processing. This is achieved with the system of tokens, which are associated with the various products and cannot be altered as they are managed on the blockchain;
- Buyers could be in charge of their products, with complete traceability of the supply chain;
- Ability to retrace the entire supply chain, simply by accessing the blockchain and public servers with documents, starting from the QR code shown on the final product.

As the IoT continues to grow rapidly, sensors and devices are becoming more and more used to communicate information. In food supply chain networks, where data such as location, temperature, humidity, or other property needs to be shared among different stakeholders, a blockchain is a natural solution for creating a tamper-evident record.

The IoT-enabled package would transmit required status information as the object under control passes through multiple carriers. The Smart Contract would specify the conditions that must be met during the shipment from the factory to the selling point (e.g., a supermarket), and all parties must adhere to the terms of the contract. For example, a humidity and temperature sensor embedded in the package stores the data and sends it to a blockchain using a dedicated IoT platform, as soon as connectivity is received. All this information would be shared among all the stakeholders involved in the process. In this way, it would be possible to check if the contractual obligations have been met throughout the supply chain. If an object moves from carrier A to carrier B, it would be possible to verify if both carriers have respected the required conditions (temperature and humidity) and how long the object has been kept in a nonprotected environment. A blockchain would allow all the stakeholders to access the same data without requiring central control.

11.5.1 A Practical Example: Cheese Production Supply-Chain

Let us consider the cheese production case study, and specifically the production of a protected designation of origin (PDO) cheese. The production process of PDO cheeses must follow precise and strict rules, to guarantee high standards and quality of the final product.

Figure 11.2 shows the stages of the production process:

1. Milking phase;
2. The milk is stored in the farm and then transported to the factory site;
3. The milk is analyzed (for quality purposes), processed, and turned into cheese;

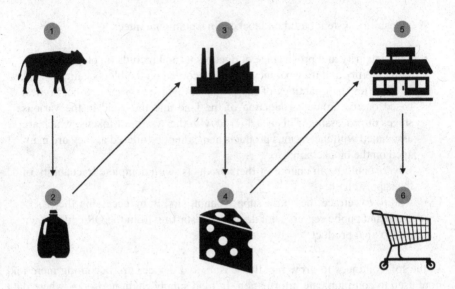

FIGURE 11.2 Stages of cheese production.

4. Cheese is stored in the factory and then transported to the shops;
5. The cheese is stored in fridges in shops;
6. Cheese is finally bought by customers.

There are several essential phases in the production process. Phase 2, e.g., after the milking process, the milk is stored in the farm and must be kept at a specific temperature to ensure a certain quality of the final product. If the temperature is too high, the milk could spoil, and therefore, it must not be used for producing the cheese. Considering an IoT packaging for the milk, the certification of the cheese supply chain could start immediately. The container could have a built-in temperature sensor connected to an IoT platform. The temperature is recorded every 10 minutes, and the value is sent and stored in a blockchain. The values of temperature recorded in the blockchain would be accessible by all the stakeholders involved in the process. For example, for phase 3 (milk arrived at the factory), the responsible person of the factory could be able to verify if the milk was stored following the procedure in the farm and during the transport phase (from the farm to the factory). This point is crucial, not only the final customer can verify the quality of all the ingredients in the supply-chain but also all the stakeholders involved.

A process with an IoT supported solution can be the following proposed in Figure 11.3:

A sensor of temperature will be able to record (and the IoT devices able to store the data in a blockchain, Figure 11.4) the temperature of the milk from stage 3, and then the temperature of the cheese will be monitored after stage 4. The final customer will be able to verify if the level of temperature of both the ingredients and the final product was kept in the correct quality range.

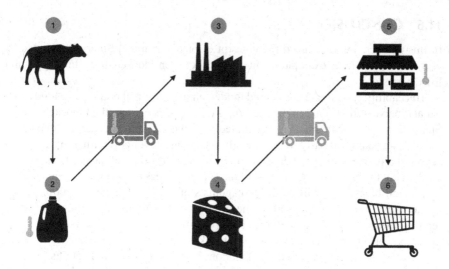

FIGURE 11.3   Internet of Things (IoT) temperature sensors.

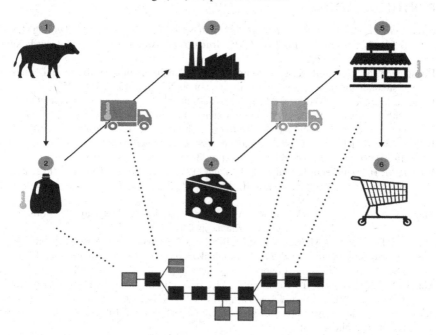

FIGURE 11.4   Solution with underlying blockchain.

Smart Contracts could be used for automatizing the entire supply chain. For example, during stage 2, only those milk containers which respected the temperature range accorded for the process will be loaded on the truck and reach the factory for the next step. Likewise, with the final product, only those pieces of cheese which were kept within the range of temperature which guarantee the quality of the product will enter the shop.

11.6 CONCLUSION

In this chapter, we introduced the concept of blockchain and Smart Contracts and provided a high-level example of integration between blockchain technologies and IoT devices.

Traceability systems are considered fundamental to ensure the safety of a food product and prevent food fraud in the food supply chain. System based on blockchain and Smart Contracts for monitoring, integrated with the IoT, allow a traceability system to occur without third party, which usually is responsible for controlling the agri-food supply chain, transparency of data, and indication of the origin of the products, as well as allowing the end customer to control the characteristics of the product purchased.

IoT and blockchain will be the protagonists of a technological revolution which has already started. Understanding the implications of the combined used of these technologies will be fundamental for both practitioners and researchers. IoT devices generate, collect, transform data, while a blockchain stores data in a tamper-resistant way, and Smart Contracts are the foundations of decentralized applications.

FURTHER READING

Casaleggio, D., Di Nicola, V., Marchesi, M., Missineo, S., Tonelli, R., "A digital voting system for the 21st century," *Euro-Par 2020: Parallel Processing Workshops*, Warsaw, Poland, vol. 12480, p. 42, 2021.

Dashkevich, N., Counsell, S., Destefanis, G., "Blockchain application for central banks: a systematic mapping study," *IEEE Access*, vol. 8, 2020, pp. 139918–139952.

Fenu, G., Marchesi, L., Marchesi, M., Tonelli, R., "The ICO phenomenon and its relationships with ethereum smart contract environment," *International Workshop on Blockchain Oriented Software Engineering (IWBOSE)*, Campobasso, Italy, pp. 26–32, 2018.

Hileman, G., Rauchs, M., "Global cryptocurrency benchmarking study," *Cambridge Centre for Alternative Finance*, vol. 33, 2017, pp. 33–113.

Kassab, M. H., DeFranco, J., Malas, T., Laplante, P., Neto, V. V. G., "Exploring Research in Blockchain for Healthcare and a Roadmap for the Future,. *IEEE Computer Architecture Letters*, vol. 1, 2019, p. 1.

Marchesi, L., Mannaro, K., Porcu, R., "Automatic generation of blockchain agri-food traceability systems," 2021. ArXiv preprint arXiv:2103.07315.

Marchesi, L., Marchesi, M., Destefanis, G., Barabino, G., Tigano, D., "Design patterns for gas optimization in ethereum," *International Workshop on Blockchain Oriented Software Engineering (IWBOSE)*, London, ON, Canada, pp. 9–15, February 2020.

Marchesi, L., Marchesi, M., Tonelli, R., "ABCDE–agile block chain DApp engineering," *Blockchain: Research and Applications*, vol. 1, nos. 1–2, 2020, p. 100002.

Nakamoto, S., Bitcoin, A., "A peer-to-peer electronic cash system," Bitcoin, 2008, p. 4. https://bitcoin.org/bitcoin.pdf.

Spink, J., Moyer, D. C., "Defining the public health threat of food fraud," *Journal of Food Science*, vol. 76, no. 9, 2011, pp. R157–R163.

12 IoT Requirements and Architecture
A Case Study

12.1 INTRODUCTION

There are various challenges associated with building requirements and architecture for Internet of Things (IoT) systems. First, it is relatively a new domain, and capturing the requirements based on the proper domain knowledge is necessary before designing and developing IoT-based systems. Most of the developers do not know the IoT domain. If a software developer and requirement analyst lack domain knowledge regarding IoT, it will be difficult for them to design and develop an effective system.

Second, when specifying the functionality for IoT applications, attention is naturally focused on concerns such as fitness of purpose, wireless interoperability, energy efficiency, and so on. Conventional requirements elicitations techniques such as domain analysis, Joint Application Development (JAD), and Quality Function Deployment (QFD) among others are usually adequate for these kinds of requirements. But in some domains, as healthcare or education, where IoT applications can be deployed, some quality requirements are probably of greater concern.

Given the increased communication and complexity of IoT technology, there is an increase in security-related concerns (Georgescu & Popescu 2015). Many of the devices used in a provision, specialized IoT will collect various data whether that surveillance is known or not (Laplante, Laplante & Voas, 2015). But why are these data being collected? Who owns the data? And where does the data go? These are questions that need to be answered by the legal profession and government entities that will oversee the deployment of IoT systems in various domains.

On the other hand, by embedding sensors into front sensor/actuator environments as well as terminal devices, an IoT network can collect rich sensor data that reflect the real-time environment conditions of the front sensor/actuator and the events/activities that are going on. Since the data is collected in the granularity of elementary event level in a 7×24 mode, the data volume is very high and the data access pattern also differs considerably from traditional business data. The related "scalability" requirements will need to be addressed. Finally, deploying IoT systems opens the doors for new quality attributes to emerge. For example, there are questions on the moral role that IoT may play in human lives, particularly concerning personal control. Applications in the IoT involve more than computers interacting with other computers. Fundamentally, the success of the IoT will depend less on how far the technologies are connected and more on the humanization of the technologies that are connected.

DOI: 10.1201/9781003027799-12

In this chapter, we present a case study for building the requirements and the architecture for an IoT-based Home Automation Management System (HAMS).

12.2 CASE STUDY DESCRIPTION: HOME AUTOMATION MANAGEMENT SYSTEM (HAMS)

Consider a company that primarily sells sensors and actuators for building automation. They have software applications that manage a network of these devices, but this constitutes a loss leader, i.e., they lose money on the software but it helps the sale of the hardware devices.

Now, the company realizes that the hardware is being commoditized, and over time, the profit margins on the sale of their hardware devices are going to shrink. To sustain their business long term, the company decides to create a new building automation system that will be profitable. They wish to accomplish this by doing two things:

1. Reduce internal development costs and
2. Expand the market.

The company wished to build a single Home Management System that will manage the different sensors and actuators. With that, the company's internal development costs can be reduced by replacing several of the existing applications with a single one, while the market expansion can be achieved by entering new and emerging geographic markets and opening new sales channels in the form of supporting not only native sensors/actuators but also those from different manufacturers.

12.3 REQUIREMENTS FOR THE HOME AUTOMATION MANAGEMENT SYSTEM (HAMS)

The first step is to clearly outline the business goals of the HAMS and their refinements; Table 12.1 restates the business goals and refines these into their corresponding engineering objectives.

TABLE 12.1

Business Goals and Engineering Objectives for HAMS

Business Goal	Engineering Objectives
Reduce internal development costs	1. Integrate the four existing management solutions into a single unified building automation system
Expand by entering *new* and *emerging* geographic markets	2. Support international languages
	3. Comply with regulations impacting life-critical systems, such as fire alarms, to operate within specific latency constraints
Open new sales channels in the form of value-added resellers (VARs/VAPs)	4. Support hardware devices from different manufacturers
	5. Support conversions of non-standard units used by the different hardware devices

From the engineering objectives, we can start defining the functions a product must support. For instance, integration (*engineering objective 1*) implies the features of existing applications to be integrated (e.g., lighting, HVAC, security, and so forth) must be supported in the new system. This may require innovative ways of displaying information in the user interface and providing fine-grained access control on who is allowed to interact with what part of the system. Supporting international languages (*engineering objective 2*) implies personalization capabilities. Regulatory policies for safety-critical parts of the system (*engineering objective 3*) would require alarm handling capabilities for situations that could cause loss of life. Supporting hardware devices from different manufacturers (*engineering objective 4*) would require dynamic configuration capabilities.

Figure 12.1 presents the functionalities of the system through a use case diagram. In this diagram, we can see that the IoT engineer intends to manage sensor/actuator systems and dynamically reconfigure them. The facilities manager intends to manage alarms generated by sensor/actuator systems that monitor a building. Alarms related to events that could cause loss of life also result in notifications to the public safety system. The system administrator intends to manage the users of the building automation system.

The use cases can be further refined into a set of functional responsibilities that must be fulfilled by the system. The following responsibilities are derived from the use cases shown in Figure 12.1:

1. The system shall send commands to a sensor/actuator device.
2. The system shall receive events from a sensor/actuator device.

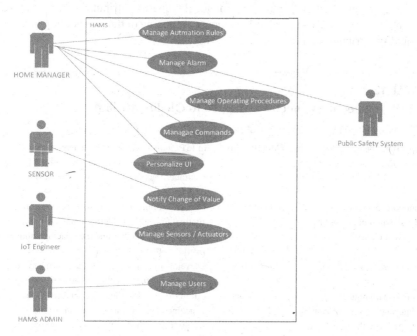

FIGURE 12.1 Use case diagram for Home Automation Management System (HAMS).

3. The system shall perform semantic translation for sensor/actuator device data.
4. The system shall route data to a sensor/actuator device.
5. The system shall evaluate and execute automation rules.
6. The system shall send automation commands.
7. The system shall generate alarm notifications.
8. The system shall display sensor data.
9. The system shall capture/relay user commands.
10. The system shall display alarm notifications.
11. The system shall edit/create automation rules.
12. The system shall retrieve data from sensor devices.
13. The system shall store sensor/actuator device configuration.
14. The system shall propagate change of value notifications.
15. The system shall authenticate and authorize users.
16. The system shall persist automation rules, user preferences, and alarms.

From the engineering objectives, we also can conclude the corresponding quality attributes the end system must exhibit. For example, to support a multitude of IoT hardware devices (*engineering objective 4*) and consider different languages (*engineering objective 2*), the system must be modifiable. To support different regulations (*engineering objective 3*) in different geographic markets, the system must respond to life-threatening events promptly (a performance requirement). It is, therefore, critical that the business goals and their implied quality concerns be fully understood.

Following the Software Engineering Institute's Quality Attribute Workshop (QAW), we can elicit concrete scenarios for the quality attributes corresponding to the engineering objectives. Table 12.2 shows a mapping of the engineering objectives to quality attribute scenarios for the HAMS.

TABLE 12.2
HAMS Engineering Objectives Mapped to Quality Attributes

Engineering Objective	Quality	Quality Scenario
Support hardware devices from many different manufacturers	Modifiability	An IoT engineer can integrate a new sensor/actuator device into the system at runtime and the system continues to operate with no downtime or side effects
Support conversions of nonstandard units used by the different devices	Modifiability	A system administrator configures the system at runtime to handle the units from a newly plugged-in sensor/actuator device and the system continues to operate with no downtime or side effects
Support international languages	Modifiability	A developer can package a version of the system with new language support in 40 person hours
Comply with regulations requiring life-critical systems to operate within specific latency constraints	Performance	A life-critical alarm should be reported to the concerned users within 60 seconds of the occurrence of the event that generated the alarm

TABLE 12.3

Constraints for HAMS

Category	Factor	Description
Organization	New market segments	Limited experience with some market segments the organization would like to enter
Technology	Scalability and responsiveness	The system must be scalable to handle a large number of sensor/actuator devices and improve responsiveness
Product	Performance and scalability	The system must handle a wide range of configurations, say, from 100 sensor/actuator devices to 500,000 sensor/actuator devices

TABLE 12.4

Architectural Drivers

	Architectural Driver	Priority
1	Support for adding new sensor/actuator device	(H, H)
2	International language support	(H, M)
3	Non-standard unit support	(H, M)
4	Latency of event propagation	(H, H)
5	Latency of alarm propagation	(H, H)
6	Load conditions	(H, H)

Table 12.3 enumerates constraints concluded for the HAMS:

From the functions, quality attribute scenarios, and constraints enumerated in the preceding sections, we distill a list of significant architectural drivers. A prioritized list of such drivers for the HAMS is shown in Table 12.4.

Architectural drivers 1 through 5 relate to the quality attribute scenarios enumerated in Table 12.2. Besides, architectural drivers 1 and 3 also correspond to the dynamic reconfiguration functions, architectural driver 2 corresponds to the internationalization and localization functions, architectural driver 4 corresponds to event management functions, and architectural driver 5 to alarm management functions. Most architectural drivers relate to the factors identified in Table 12.3. For instance, the organizational factor concerning new market segments is reflected in architectural drivers 1 through 5. These drivers take into account the flexibility needed to accommodate new sensor/actuator devices and their calibration, language and cultural aspects, and regulatory concerns regarding the responsiveness of the system to safety-critical events. The technological factor related to scalability and responsiveness and the product factor related to performance and scalability are addressed through architectural drivers 4–6.

12.4   ARCHITECTURAL OPTIONS FOR HAMS

We begin with one of our highest priority architectural driver #1 (support for adding a new sensor/actuator device). This driver is related to modifiability quality, and we need to apply modifiability tactics to limit the impact of change and minimize the number of dependencies on the part of the system responsible for integrating new hardware devices. There are three design concerns related to modifiability:

- *Localize Changes*: This relates to adding a new sensor/actuator device
- *Prevention of Ripple Effects*: This relates to minimizing the number of modules affected as a result of adding a new sensor/actuator device
- *Defer Binding Time*: This relates to the time when a new sensor/actuator device is deployed and the ability of nonprogrammers to manage such deployment

We address these concerns by creating IoT adaptors for sensor/actuator devices, "an anticipation of expected changes" tactic. We use two additional architectural tactics to minimize the propagation of change. First, we specify a standard interface to be exposed by all IoT adaptors to "maintain existing interfaces". Second, we use the IoT adaptor as an "intermediary" responsible for the semantic translation into a standard format, of all the data received from different sensor/actuator devices. As a side effect, this also addresses architectural driver #3.

The adaptors are assigned the following responsibilities:

- Send commands to the sensor/actuator device
- Receive events from sensor/actuator device
- Perform semantic translation for sensor/actuator device data

Instantiating and allocating responsibilities to the adaptors leads to a realization that the building automation server is still sensitive to a change in the number of sensor/actuator devices it is connected to and must include logic to route commands and data, to and from the correct adaptor. To address this concern, we use the "hiding information" tactic introducing an IoT adaptor manager to hide information about the number and type of sensor/actuator devices connected. The IoT adaptor manager together with the adaptors creates a virtual device, i.e., for all other components of the building automation system, there is practically one sensor/actuator device to interact with at all times.

Additionally, the IoT adaptor manager uses the following two architectural tactics to address the defer binding time design concern:

- "Runtime registration": This will support plug-and-play operation allowing nonprogrammers to deploy new sensor/actuator systems
- "Configuration files": This tactic enables the setting of configuration parameters (such as initial property values) for the sensor/actuator systems at startup

The IoT adaptor manager is assigned the following responsibilities:

- Configure a sensor/actuator device
- Route data to a sensor/actuator device

At this stage, architecture driver #1 is satisfied (and as a side effect so is architecture driver #3), we next consider architectural drivers 4– 6 related to the performance quality attribute of the HAMS. There are two design concerns related to these drivers:

- *Resource Demand*: The arrival of change of property value events from the various sensor/actuator devices and the evaluation of automation rules in response to these events are a source of resource demand
- *Resource Management*: The demand on resources may have to manage to reduce the latency of event and alarm propagation

To address these concerns, we move the responsibility of rule evaluation and execution, and alarm generation, respectively, to a separate Logic & Reaction (L&R) component and an alarm component. These components running outside the automation server can now be easily moved to dedicated execution nodes if necessary. In doing so, we are making use of the increased available resources tactic to address the resource management concern and the reduced computational overhead tactic to address the resource demand concern.

We use an additional tactic to address the resource management concern. This tactic relies on introducing "concurrency" to reduce delays in processing time. Concurrency is used inside the L&R and alarm components to perform simultaneous rule evaluations.

We assign the following responsibilities to the L&R and alarm components:

- Evaluate and execute automation rules
- Send automation commands

Besides, we assign the following responsibility to the alarm component:
- Generate alarm notifications

The results of applying these tactics are shown in Figure 12.2.

We finally consider architectural driver 2 related to international language support for the HAMS. There are two design concerns for this driver:

- *Localize Changes*: This relates to changing the user interface to deal with a new language and culture
- *Prevention of Ripple Effects*: This relates to minimizing the number of modules affected as a result of changing the user interface

We address these concerns using the following modifiability tactics to address localize changes and prevention of ripple effects design concerns:

- *Anticipation of Expected Changes*: Changes to the user interface (UI) are localized to the presentation component
- *Intermediary*: The presentation component acts as an intermediary preventing ripple effects from changes to the UI from propagating to the rest of the application.

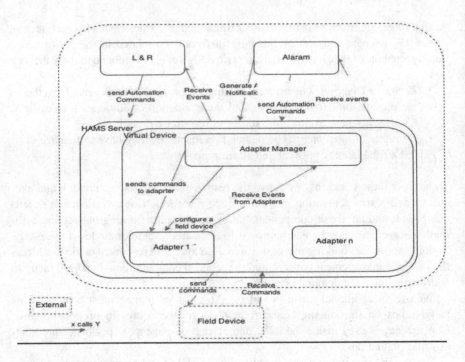

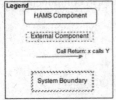

FIGURE 12.2 Home Automation Management System (HAMS) architecture after applying architectural drivers 1, 4, 5, and 6.

We assign the following responsibilities to the presentation component:

- Display device data
- Capture/relay user commands
- Display alarm conditions
- Edit/create automation rules

The results of applying these tactics are shown in Figure 12.3.

The architecture elaboration process we are using is iterative. Moreover, the tactics we choose to implement can very often have a negative impact on the quality attributes they do not target specifically. In the case of the building automation system, we focused on modifiability and performance tactics which can have a negative impact on each other. We revisit the performance and modifiability drivers next to address these issues.

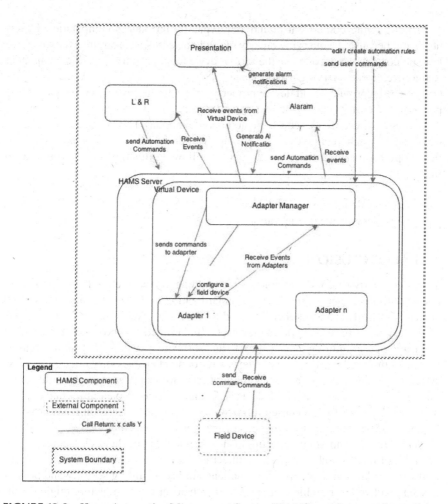

FIGURE 12.3 Home Automation Management System (HAMS) architecture after applying all architectural drivers.

- Introducing performance tactics resulted in the creation of multiple components (Rule Manager and Alarm Manager, for instance) that now depend on the virtual device. Therefore, based on its current structure, we can predict that some changes to the virtual device have the potential to propagate to several other components of the building automation system. We would like to minimize the ripple effect of these changes. To achieve this objective, we introduce a Publish-Subscribe bus. We allocate the following responsibility to the Publish-Subscribe bus:
- Propagate change of value notifications

By examining the current system structure, it can be seen that every time the Rule Manager and the Alarm Manager needs to query a field device, it needs to make a call that traverses multiple components along the way to the field device.

Since crossing component boundaries typically introduces computational overhead, and because the querying latency of field devices is a constraint over which we have no control, we decompose the virtual device and introduce a cache component to improve device querying performance.

This cache provides field device property values to the system, saving part of the performance cost incurred when querying the actual field devices. The performance gains are seen because we reduce the number of component and machine boundaries traversed for each query. A cache is an application of *maintaining multiple copies of data* performance tactics. We allocate the following functional responsibilities to the cache:

- Retrieve data from a field device
- Store field device configuration

12.5 CONCLUSION

The IoT is relatively a new technology, and building an architecture for an IoT system comes with challenges. The lack of standardization is one of the significant challenges. The hardware components related to the IoT as well as the software aspects of the IoT don't have a single platform of standardization. Having a standardized application programming interface (API) and software service so that future applications are employed in a uniform and level environment enables easy migration across systems. The interoperability of another concern. IoT applications offer value combining data sets from various IoT devices. Connectivity is also an important component of IoT architecture because it plays a crucial role in transporting data from the sensors. Furthermore, it also transmits instructions to the actuators.

While there is no single and uniform agreement about the architecture of IoT, according to some architects, IoT architecture has three layers. While others, support the four-layer architecture. The challenge in IoT regarding security and privacy has led to a five-layer architecture proposal. Whatever the architecture might be, the challenges of an IoT architecture almost remain the same.

FURTHER READING

Georgescu, Mircea, and Daniela Popescu. "HOW COULD INTERNET OF THINGS CHANGE THE E-LEARNING ENVIRONMENT." *eLearning & Software for Education* 1 (2015).
Laplante, P., N. Laplante, and J. Voas. "Considerations for healthcare applications in the internet of things." *Reliability* 61.4 (2015): 8-9.

Index

Note: **Bold** page numbers refer to tables and *italic* page-numbers refer to figures.

Printed in the United States
by Baker & Taylor Publisher Services